Visualizing Dynamic Systems:
Volumetric and Holographic Display

Synthesis Lectures on Engineering, Science, and Technology

Each book in the series is written by a well known expert in the field. Most titles cover subjects such as professional development, education, and study skills, as well as basic introductory undergraduate material and other topics appropriate for a broader and less technical audience. In addition, the series includes several titles written on very specific topics not covered elsewhere in the Synthesis Digital Library.

The Big Picture: The Universe in Five S.T.E.P.S.
John Beaver
2020

Relativistic Classical Mechanics and Electrodynamics
Martin Land and Lawrence P. Horwitz
2019

Generating Functions in Engineering and the Applied Sciences
Rajan Chattamvelli and Ramalingam Shanmugam
2019

Transformative Teaching: A Collection of Stories of Engineering Faculty's Pedagogical Journeys
Nadia Kellam, Brooke Coley, and Audrey Boklage
2019

Ancient Hindu Science: Its Transmission and Impact on World Cultures
Alok Kumar
2019

Value Rational Engineering
Shuichi Fukuda
2018

Strategic Cost Fundamentals: for Designers, Engineers, Technologists, Estimators,
Project Managers, and Financial Analysts
Robert C. Creese
2018

Concise Introduction to Cement Chemistry and Manufacturing
Tadele Assefa Aragaw
2018

Data Mining and Market Intelligence: Implications for Decision Making
Mustapha Akinkunmi
2018

Empowering Professional Teaching in Engineering: Sustaining the Scholarship of Teaching
John Heywood
2018

Visualizing Dynamic Systems: Volumetric and Holographic Display
Mojgan M. Haghanikar

ISBN: 978-3-00964-8 print
ISBN: 978-3-02092-6 ebook
ISBN: 978-031-00164-2 hardcover

DOI 10.1007/978-3031-02092-6

A Publication in the Springer series Publishers series
SYNTHESIS LECTURES ON ENGINEERING, SCIENCE, AND TECHNOLOGY
Lecture #15

Series ISSN 2690-0300 Print 2690-0327 Electronic

Visualizing Dynamic Systems:
Volumetric and Holographic Display

Mojgan M. Haghanikar
SETI Institute

SYNTHESIS LECTURES ON ENGINEERING, SCIENCE, AND TECHNOLOGY #15

ABSTRACT

This book is aimed to help instructional designers, science game designers, science faculty, lab designers, and content developers in designing interactive learning experiences using emerging technologies and cyberlearning. The proposed solutions are for undergraduate and graduate scientific communication, engineering courses, scientific research communication, and workforce training.

Reviewing across the science education literature reveals various aspects of unresolved challenges or inabilities in the visualization of scientific concepts. Visuospatial thinking is the fundamental part of learning sciences; however, promoting spatial thinking has not been emphasized enough in the educational system (Hegarty, 2014)[1]. Cognitive scientists distinguish between the multiple aspects of spatial ability and stresse that various problems or disciplines require different types of spatial skills. For example, the spatial ability to visualize anatomy cross-sections is significantly associated with mental rotation skills. The same is true for physical problems that often deal with spatial representations. However, most of the physics problems are marked by dynamicity, and visualizing dynamicity is inferred by the integrations of different participating components in the system. Therefore, what is needed for learning dynamicity is visualizing the mental animation of static episodes.

This book is a leap into designing framework for using mixed reality (XR) technologies and cyberlearning in communicating advanced scientific concepts. The intention is to flesh out the cognitive infrastructure and visuospatial demands of complex systems and compare them in various contexts and disciplines. The practical implementation of emerging technology can be achieved by foreseeing each XR technology's affordances and mapping those out to the cognitive infrastructure and visuospatial demands of the content under development.

KEYWORDS

complex systems, visualization, mixed reality, spatial ability, science education, cyberlearning

---

[1]   M. Hegarty, "Spatial thinking in undergraduate science education," *Spatial Cognition and Computation*, 14, 142–167, 2014

I dedicate this book to all teachers and instructors around the world for their steadfast diligence, especially during the pandemic!

Contents

Acknowledgments

The completion of this book would not have been possible without the support, encouragement, and faith that my family, professors, and friends offered. My deepest gratitude is due to the members of AWE (Augmented Reality World Expo), NYC Lab for arranging informative conferences, meetups, and workshops, as well as the AAPT Committee on Educational Technology for their support and for promoting the use of technology to advance inquiry. Sincere gratitude goes to Dr. Herbert Fotso for illuminating conversations on creating interactive renderings for quantum computation. Also, I am thankful to artists, photographers, physicists, and mathematicians who allowed me to use their mind-blowing creations. Professor Robert Devaney, Professor Wolfang Beyer, Tiffany McFarlane, Paul Nylander, and Ingo Berg generously offered their sophisticated fractal pictures. Thanks to Alex Koloskov for the professional high-speed photography on colorful liquid arts. Special gratitude to zSpace, Refik Studio, Reflect Institute, Space Telescope Science Institute, NASA, Labster, and Karen Vanderpool-Haerle for sharing their images. Alexander Teverovsky's research results from his fellowship at NASA Space Technology were a great addition to the discussion of dendrites, and not forgetting my friends Anita Stuckensmith and Maciej Czarnecki, who shared their artwork of snow dendrites and a time-lapse video of clouds. I wish to express my heartfelt gratitude to Professor Dean Zollman for his continuous support of my scholarship activities and Professor Sorenson for his unending fresh ideas in physics and stimulating discussions on developing capstone projects. Lastly, I am thankful for my sister Taraneh M. Haghanikar who provided endless support during the pandemic. Her continued support and encouragement made all the difference.

Introduction

This book is a leap into designing techniques that uncovers the invisible paradigm of scientific interactions. For the first time, the frontiers of technology such as mixed realities have provided means for communicating dynamicity of scientific concepts as of the complex systems. The aim is to provide guiding principles blended from educational research, Human–Computer Interactions (HCI) and EdTech solutions to stay ahead of design practices. The targeted audiences are educators, instructional designers, science educators, science teachers, science lab instructors, cognitive scientists, UX designers, professionals across industries, workforce trainers, science and engineering instructors and researchers, content developers, andgame designers. If you are not a designer but a decision maker on an educational platform, this book provides a diligent guide for how to communicate with your designers and consultants in your team.

Before discussing the implications of emerging technologies such as Virtual Reality (VR), and Augmented Reality (AR) in scientific communication, we need a framework to start. This framework should classify various subjects' cognitive demands and find an appropriate technology to create an EdTech-assisted rendering. To this end, the first few chapters deal with a review of complex systems, the distinction of various dynamic processes and system thinking necessities, essential skills such as visuospatial skills, and a review of science education literature on challenges in visualization. The aim is to search for a classification scheme to categorize various complex systems in terms of their system thinking and spatial demand.

Design with mixed realities is cost-effective and to make the best of the design requires understanding rules in this new frontier of extended reality. A crucial part of achieving a good design is clearly understanding the objectives and users' needs and capacities. Like a filmmaker whose big shoot is coming up, you may be excited about every scene you want to create. Like making film projects that storyboard sketches the scenes, storyboarding for mixed realties is visualizing every scene of your rendering design. However, storyboarding for scientific communication is not about sharing experiences and knowledge but about the fine skills of the audience. A good storyboard needs a specific framework tailored to the audience, and that is not possible unless taking into account the previous research and evaluation that should be served as evidence-based strategies. Ultimately, a successful design is defined by an optimum balance between all the parameters involved. This book is a roadmap to bridge the previous research and experiences to a promising design with mixed reality technologies.

Overview of Chapters

CHAPTER 1: THE ART OF THINKING ABOUT COMPLEX SYSTEMS

Telescopes are a tool for the naked eye to uncover the farthest things in the sky, and microscopes reveal earthly items too small to see. Nevertheless, what are the tools needed for naked perception to uncover the behavior of nature and the invisible interactions? The ultimate goal of learning is to understand better the world and nature, which are systems with different complexity levels. The complexity is a challenge, yet our thinking is often biased and simplistic. Scientific models are tools and mental representations of natural phenomena. Comprehending the complex and nonlinear world requires a system thinking that is up to the mark with the complex system's cognitive demand. Accordingly, every new type of system is like starting a new paradigm that requires suitable and unique toolkits. In this view, various styles of thinking about systems are classified, including static, dynamic, deterministic dynamic, and chaotic. This classification scheme can be used to categorize various types of complex systems in terms of their system thinking and spatial demand.

CHAPTER 2: SPATIAL ABILITIES AND SUCCESS IN SCIENCES

Deeping dive into different aspects of visualization in sciences, this chapter provides an overview of psychometric cognitive science literature to distinguish the types of spatial abilities and highlight the visuospatial requirements central to visualization to various disciplines. We will provide an overview of cognitive studies to understand better students' challenges in visualizing spatial representations and subsequently suggest resolutions to nurture spatial thinking in sciences using emerging interactive technologies.

CHAPTER 3: SCIENCE EDUCATION LITERATURE: INVISIBLE CONCEPTS

Visualizing abstract concepts and processes is one of the most critical challenges of learning physics. This chapter aims to present an overview of visualization challenges that have been reported in physics education. The discussion is on many types of invisible concepts in sciences, comparison of mental visualization of concepts over the different time and scale, and mental visualization of dynamic processes.

CHAPTER 4: EDTECH SOLUTIONS

EdTech is a practice of implementing evidence-based teaching principles by using technology. The evidence-based practices rest on four pillars of cognitivism, constructivism, social constructivism, and visualization. Education designers use a technology genre to promote student-centered learning experiences. Student-centered practices allow users to delve into real-world scenarios and test their problem-solving skills, teamwork, communication, and collaborating abilities.

However, choosing a helpful EdTech might be like searching for a needle in a haystack. From numerous EdTech productions how can a user select a credible intervention? How may the user know if the voice of the educator was incorporated in the design? Users should have access to a credible source that provides vetted resources for a diverse set of users.

comPADRE[2] is a network of accessible online resource collections supporting faculty, students, and teachers in physics and astronomy education. The other collections of resources are PhysPort[3], which inventorized the resources developed by physics education researchers and collected the use of numerical modeling at all levels of physics education, and PICUP, a community for those promoting computation in the physics curriculum (Mason, 2012).

This chapter revisits the decades of using EdTech solutions in science. The future designs using emerging technologies can be built on previous trends. Such trends provide insight into educational designers' minds and help them fix the challenges previously faced by both students and educators.

CHAPTER 5: EMERGING TECHNOLOGIES: A TWIST IN EDTECH SOLUTIONS

The recent advancements in mixed reality developments hold significant potential to unlock visualization in three dimensions, eliminating the barriers of experimentation and interaction in e-learning. In addition to 3D allowance, mixed reality technologies enable interactions with digital objects merged into the physical world. This chapter looks at the attributes of these technologies and some current examples of how they are being used and comparing their hardware, software, and affordances that suit various purposes. The multiple facets of VR are discussed and differentiated from AR that blends the digital renderings projected over physical surroundings within people's field of vision. The discussion will follow Hololens or zSpace interactions in which an expanded field of view extends out of the screen to the surroundings and allows more possibilities for comparing and exploring the cause-and-effect relationships.

---

[2] https://www.compadre.org.
[3] https://www.physport.org.

CHAPTER 6: CURRICULUM DESIGN AND EMERGING TECHNOLOGIES

The emergence of Massive Online Open Courses (MOOCS) was a game-changer in education which started with online streaming by professors at Stanford University. However, there are many aspects of constructivism that cannot be implemented employing remote courses' current means. In this view, Chapter 6 introduces a new generation of virtual science labs to present the particular affordances of cyberlearning with different resolutions to sketch a unique platform for communicating science and engineering concepts. In doing so, this chapter offers insights on selecting the type of emerging technology according to a lesson design's purpose and objectives.

CHAPTER 7: BREAKTHROUGHS IN SCIENTIFIC COMMUNICATION

Like every history book that rests upon fierce rivalry that shaped the revolution's course, this chapter discusses how the previous research of learning challenges informed our new designs with the technology genre. Discussing the challenges facing scientific communication and learning from previous developments are laying a launchpad with varying preparedness levels for fully understanding new potentials and breakthroughs. The deterioration of traditional instruction methods has set the stage for implementing a complicated set of integrated designs.

The critical attributes of the AR/VR technologies have made a breakthrough in remote science labs that could replace e-learning with cyberlearning, provide virtual benches for workforce training, and a new platform that reshapes the research and assessment.

CHAPTER 1

The Art of Thinking About Complex Systems

1.1 INTRODUCTION

The quest of science is to explore how nature works. Many aspects of nature are observables, but many other aspects are invisibles, and the only way to detect the invisible is to trace their interactions with the observables through a chain of cause and effect. The invisible aspects that escape our senses are often too far, small, fast, or slow. The physical entities are not always substances; they are also in the forms of waves and energies. Maybe we can see the standing waves in a string, but in quantum mechanics, the quantum state of a particle, which is called wave function, is invisible and defined by the probability of locating the particle at that point. We have a limited range of sensibility to detect the spectrum of waves. The other type of invisible is a pile of data. The most invisible aspect of nature is processing too much information. Through the ages, humans have invented many instruments to compensate for their limitations to observe and explore. Telescopes are designed to observe the far, microscopes for the invisible tiny structures, and computers for processing and recording bulky data. With the aid of proper instrumentations, invisible aspects of nature can be discovered, recorded, and represented. This book aims to discuss a particular type of invisibles that are not in the form of substances but in the form of patterns, interactions, interrelationships, and behaviors. The main goal is to introduce suitable instrumentations for capturing the most hidden part of nature and its complex system: the *interrelationships, interactions, links between cause and effect, patterns*, and *processes*.

Scientists often face a variety of perplexing challenges dealing with complex systems. The study of complex phenomena requires competency, efficient communication and collaboration across the disciplines, and interdisciplinary training. The study of natural phenomenon necessitates an interdisciplinary approach. In order to bridge the gap between cross-border research and elucidate barriers of communications, interdisciplinary teaching is often encouraged in undergraduate science (Nae, 2017; Buchbinder et al., 2005)—meaningful learning associates with the reconstruction of concepts in a new context or discipline (Bransford et al., 2000). Understanding the inherent complexity of natural phenomena draws on various disciplines and, consequently, on various representation tools and integrative research approaches. Traditional practices tend to compartmentalize the disciplines, and subsequently, the experts in each field are equipped with compartmentalized knowledge, and compartmentalized spatial and cognitive skills pertain to constructing that type of knowledge. As a general rule, is needed to distinct cognitive infrastructure hierarchies and visuospatial demands of complex systems and compare the demands that existed for various types of complex systems and inter-disciplines.

There are many examples of complex systems in nature, society, environment, climate, or living organisms. The human body, our learning activities, cell of a living organism, colony of ants and beehives, flock of birds, transmission and spread of infectious diseases, climate change, and social behaviors are examples of complex systems. Frequently, the constituents and building blocks of complex systems are also complex organisms or systems. Many processes in nature, social context, or artificially engineered can be categorized as complex systems. One way of describing the complex system is to model the system's emergent collective behavior as a whole and discern the collective performance from the function of system entities.

Complex systems in general are marked with patterns of commonalities such as nonlinearity and unpredictability. Systems can be studied in terms of their composite elements' interrelationships, and when systems are not complex, the emergent behavior can be predicted based on those interrelationships. As a result, a simple system's behavior can be modeled by the participating entities and their interrelationships, whereas in a complex system, a more significant number of interacting entities contribute to the emergent behavior, and even different systems analyzing tool or data collecting methods apply to complex systems. The superimposed interrelations of the participating entities create a convoluted outcome and not always linear and predictable as it can be affected significantly by the small random behavior of entities and the initial conditions.

Science is studying natural phenomena and understanding their functionality by seeking models that capture natural phenomena' complexity and dynamicity. To describe a system, scientific models often leverage mathematical and computational tools to simulate the system's mechanism into a new representation based on mathematical and computational algorithms. Scientists often use mathematical models to describe the data attained through observations and experimentations to perceive the dynamics and architecture. The models are mathematical tools or graphical representational tools that display interrelationships in a compact illustration with distributed information to ease mental visualization.

Numerous scientific theories have been employed to describe natural and artificial complex systems in various disciplines. However, there have been commonalities in research methodologies, modeling, and mathematical formulation. Wolfram (2020) sketched an outline for the study of complex systems to derive mathematical models that capture the essential features of the collected complex behavior generated due to many components interacting dynamically. Wolfram noted that many systems' essential components are simple but capturing many constituents' interaction transcends contemporary mathematics limits.

1.2   DYNAMIC SYSTEMS: UNVEILING THE INVISIBLE PATTERNS

Understanding a phenomenon involves constructing an internal cognitive structure that is a mental representation of an object or process, and there is a large number of possible cognitive structures that can represent a new stimulus (Sedikidies and Skoronski, 1991). As the scientific knowledge evolved through the history of science, more advanced and abstract cognitive representations emerged subsequently to model the new stimulus. Studying a dynamic system involves dynamic thinking, which requires an exemplary infrastructure of cognitive and visuospatial skills. Dynamic thinking incorporates the elements of time (Bratianu, 2007) and should be distinguished from inertial thinking that is the habit of novices and associates with static objects. Dynamic systems involve processes in motion. However, not all dynamic systems can be deterministic.

Figure 1.1: Analemma Image of Analemma by Giuseppe Donatiello is licensed under Creative Commons CC0 1.0. https://commons.wikimedia.org/wiki/File:Analemma_A14_2016_(25907420783).jpg.

For example, analemma or the "Sun's figure eight" is a diagram showing the position of the Sun in the sky as seen from a fixed location on Earth at the same mean solar time, as that position varies over a year. If the Earth's orbit around the Sun was a complete circle, the Sun would rise and set on the same path across the sky, always with the same altitude. In other words, if the Earth's axis wasn't tilted from the plane of its orbit around the Sun, the Sun's altitude would not form an analemma over a year. However, the Earth's orbit is elliptical, and the Earth's axis is tilted by roughly 23.5° to the orbit. The combination effect of earth's orbit and tilted axis makes the analemma.

Tracing the Sun's altitude that shapes an analemma represents a deterministic process, as replication yields the same pattern. In real time, all an observer can see is the Sun's altitude going up and down, but the only time the hidden cyclic pattern appears is after recording the Sun's location over a year from the exact location same time. In this way, a model is generated representing the cyclic variations of the altitude of the Sun. Visualizing the system of the Earth–Sun orbit required a cognitive representation of a 3D dynamic geometrical model and the Sun's altitude over the span of year can be defined by a differential equation.

In contrast, one can find many dynamic processes which are not deterministic. A straightforward example is the pattern of clouds that is constantly changing. I used to live in NYC and watching the sky's different patterns at sunset was always a surprise.

Figure 1.2: Sunset NYC. Courtesy: author.

For example, replicating the recording of a video time-lapse of the motion of clouds from a fixed location such as a window will differ each time as the form of clouds, and their motions and the shadows they cast on the buildings, roofs, or ground can be chaotic and unpredictable. Although the details about the motion of the clouds are chaotic, some features remain unchanged, and this is how we can distinguish a cloud.

Figure 1.3: Time-lapse video of the clouds. Reproduced with permission from: Maciej Czarnecki. Personal communication.

1.3   ARTIFICIAL INTELLIGENCE AND 3D TIME-LAPSE REPRESENTATIONS

The real essence of active learning is promoting scientific reasoning to describe a phenomenon; therefore, data acquisition and data interpretation are the core subset of inquiry. According to Symons and Boschetti (2013), the underlying patterns of dynamic systems remained hidden when an extensive tabulation of data obscure them. Instead, graphical representation of data such as charts and diagrams, the trends, patterns, and interrelationships are toolkits to unmask the hidden patterns. Infographics compacts the data and highlights the trends and facilitate the communication of research outcomes.

Machine hallucinations are the latest in watching 3D time-lapse of the large number of images taken over the span of time. Google satellite, biological and astronomical imaging advancements, and many other archiving improvements have provided access to large image databases. A great innovative artist, Anadol (2020), blended art and AI and machine learning to create a machine hallucination that is a dynamic time-lapse of 100 million photographs into 3D movies projected in large halls and theatres. Machine Hallucination was an exploration of time and space experienced through New York City's public photographic archives.

When Refik Anadol heard about the Alzheimer's disease of his beloved uncle, his coping mechanism was to remember memories in a new light. According to his TED talk (Anadol, 2020), he was determined to bring memories into life, so they were not disappearing, but they were re-shaping. Therefore, he invented an immersive data visualization technique that can be applied to any big dataset distributed in the span of time and space.

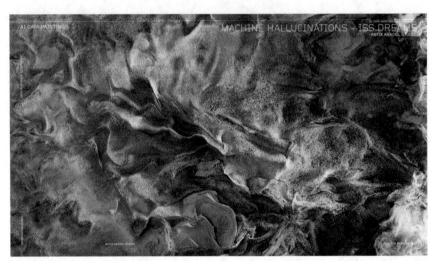

Figure 1.4: Machine hallucination. Reproduced with permission from: Refik Anadol Studio.

Figure 1.5: Machine Hallucination Exhibition. Reproduced with permission from: Refik Anadol Studio, https://refikanadol.com/works/machine-hallucination/.

The upshot of all these is the advancement of artificial intelligence (AI) and archiving, which have offered flexibility in tracking dynamicity in extended periods. Therefore, lengthy observations of the chaotic systems may shed new light on the chaotic behavior. Comparing the above examples, it is evident that chaotic dynamic systems do not have a deterministic pattern as it might be found for analemma but may show a pattern that can be found similar to the Mandelbrot set, which is coming next.

1.3.1   MANDELBROT SET

Benoit Mandelbrot (1982) saw a pattern for deterministic chaotic systems. In the same fashion of tracking the sun's location, he tracked the iterations of quadratic functions. He considered himself a fractal expert and was inspired by self-similarities in nature. His interest brought him to the Julia sets, which consist of values such that an arbitrarily small perturbation can cause drastic changes in the sequence of iterated function values.

The simplest definition of Julia set has been provided by McGoodwin (2000), who presented Julia sets as the family of sets generated by the special quadratic case form $f(z) = z^2 + c$. Here z represents a complex number that can take on all values in the complex plane. The constant "c" is also defined as a complex number, but it is held constant for any given Julia set. Therefore, there is an infinite number of Julia sets, each defined for a given value of c.

By iterating $f(z)$ values as the number of iterations increases toward infinity, either $f(z)$ can continue to grow and blow up or stay bounded. There are two group of points in the complex plane. Those that do not stay bounded with successive iterations called escape set and other points in the complex plane that stays bounded called prisoner set as many iterations are taken to infinity. All points must either be in one or the other set. The common boundary between the escape set and the prisoner set is called the Julia set.

Mandelbrot's addition to Julia's sets was to visualize the patterns of chaotic systems with the technology he had access to in 1978. As he worked for IBM, he came up with the idea to use computers to obtain the graphical representations of chaotic behavior. He graphed the iterations of quadratic functions.

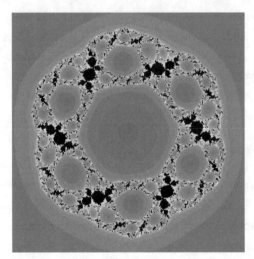

Figure 1.6: The connected Julia set for the map $z^3 + (.105 + .05i)/z^2$. Reproduced with permission from: Dr. Robert Devaney.

Figure 1.7: Fractal Set: Inside the hive. Reproduced with permission from: Tiffany Mcfarlane, https://www.trmdesignco.com/inside-the-hive.html.

Bear in mind the invisibility that was unmasked with Mandelbrot's innovative strategy. His work's significance was not just demonstrating the hidden patterns embedded within the chaotic iterations of quaratic equations, but a method that can be applied to any factes of chaotic systems in nature. Therefore, within the apparent randomness of complexity and chaos there are invisible patterns, interconnectedness, constant feedback loops, repetition, self-similarity, fractals, and self-organization which can also be considered as deterministic behavior. Chaos theory is often associated with many variable systems. Nevertheless, due to the 20th century's prominent discoveries on chaos and dynamic systems, the simplest type of systems can also exhibit chaotic behavior (Devaney, 2018). Scientists have begun studying simple systems' chaotic behavior to build a foundation for further research in turbulent many variable systems. Using this approach, Devaney (2018) has made chaos theory accessible to the learner by illustrating the chaotic behavior in normal and straightforward quadratic functions. Quadratic equations are easy to be solved and graphed. To add dynamicity to a quadratic equation, the function is needed to be composed upon itself, which is called an iterative process. To iterate a function, we should take the output of a function and feed it as an input to repeat the operation repeatedly (Devaney, 2018). In doing so, for some types of functions, a set of input numbers grouped as Julia sets could be found, resulting in totally bizarre, random outcomes and susceptibility to initial conditions, whereas the behaviors of other groups of numbers are predictable. The set of input numbers with the bizarre outcome that was grouped as escape set and Julia set includes values that with an arbitrarily small perturbation, a drastic change can be obtained in the sequence of iterated function values. Julia sets are an example of fractals as they create a self-similar subset over and over again. However, the patterns of chaotic systems are prone to drastic change with any minor changes of initial conditions that may substantially change the overall outcome.

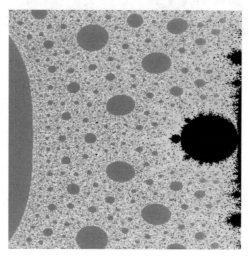

Figure 1.8: A magnification of the parameter plane for the family of maps $z^2 + C/z^2$. Reproduced with permission from: Dr. Robert Devaney.

1.4   CHAOS AND ENERGY DISSIPATION

Open systems that exchange energy with the environment can change the systems' state and initial conditions. One visually good example is a magnetic pendulum that shows the chaotic behavior under three magnets and gravitational field (Berg, 2020). The experiment was originally conducted by German scientists Hilgenfeldt and Schulz (1994). The system constitutes a magnetic pendulum sways by the net forces of gravity and magnets located on an equilateral triangle's edges. The pendulum is suspended from a pivot that is located on a vertical axis passing through the center of equilateral triangle. The pendulum's path's chaotic patterns appear after simulating the motion of a magnetic pendulum under the influence of three magnets and gravity. A color-coded pattern is obtained by tracking the pendulum assigning three colors to each magnet and recording the color of the magnet over which the pendulum came to a rest.

The pattern is dynamic, and any minor variation in the pendulum's starting point will result in a vastly different color-coded pattern. The friction cause dissipation and change of initial conditions. Like every chaotic system, the simulation results are susceptible to slight variations in their initial conditions, including the pendulum's length.

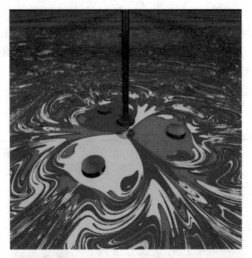

Figure 1.9: Chaotic magnetic pendulum. Reproduced with permission from: Paul Nylander (http://bugman123.com).

Figures 1.10, 1.11, 1.12, and 1.13 are several examples of various patterns obtained as a result of changing the variables such as length of pendulum or gravitation.

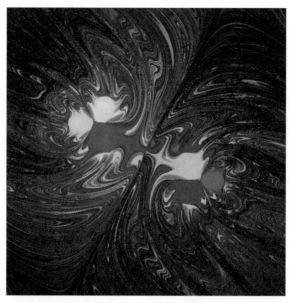

Figure 1.10: Magnetic pendulum with change of variables. Reproduced with permission from: Ingo Berg.

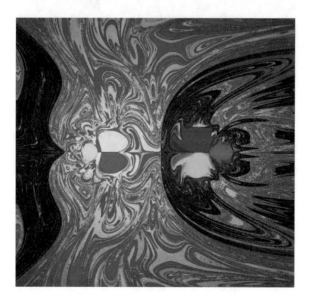

Figure 1.11: Magnetic pendulum with change of variables. Reproduced with permission from: Ingo Berg.

Figure 1.12: Magnetic pendulum with change of variables. Reproduced with permission from: Ingo Berg.

Figure 1.13: Magnetic pendulum with change of variables. Reproduced with permission from: Ingo Berg.

1.5   CHAOS IN NATURE

As discussed earlier, dynamic systems are the type of systems whose characteristics change with time, and their mathematical representations are time variant. The time-dependent functions are either differential equations or iterated functions (Feldman, 2019). Differential equations play a significant role in modeling physical sciences problems. Nevertheless, what type of behaviors represent iterated functions? Complex systems with chaotic behavior and iterated modeling also have their commonalities and distinctions. Below are examples of chaotic behavior from various domains with commonalities in mathematical modeling, including turbulent flow, crystal growth, aggregations and supramolecules, ecological and social systems. My point here is not checking out the example from a scientific view but emphasizing each's hidden aspects through cognitive lenses.

1.5.1   TURBULENT FLOW

The chaotic motion of gas flow or liquid fluids are examples of a complex system. Fluid motion can be laminar or turbulent. As one of the paragons of chaotic systems, turbulent flow can be found everywhere. There is a turbulent flow when water starts boiling in a pot until rhythmic ripples start to appear or when the draining water in the bathtub start swirling.

Figure 1.14: Imaging candle flow with Schlieren optics set up. Reproduced with permission from: https://commons.wikimedia.org/wiki/File:Laminar-turbulent_transition.jpg.

Either of two types of flow may occur depending on the fluid's velocity, density, and viscosity: laminar flow or turbulent flow. Looking at the rising smoke from a cigarette illustrates the difference between the chaotic and laminar flow. Laminar is an orderly flow that is associated with lower velocities, and turbulent flow is a less orderly flow. The flow patterns can be predicted by Reynold's number which is a dimensionless number determined by the ratio of inertial forces and resistant forces to deformation.

A Schlieren optics set up can be used to make visible the warm convection currents rising from a candle flame or cold air sinking from a glass of ice water. A turbulent flow motion is an unsteady flow and its flow properties throughout the location of the flow change with time.

Atcheson (2007) suggested an exciting method to reconstruct the 3D flow by capturing multiple viewpoints with Schlieren mirrors then blending the images topographically to reconstruct 3D models of the fluid's temperature distribution. To model the flow, particle image velocimetry (PIC) can be used. In this method, tiny colorful seeds can be traced, which are injected into the flow.

Put differently, turbulent flow represents chaotic behavior in terms of changes of its characteristics such as pressure, flow velocity, and concentration. For example, the mathematical model of blood flow on blood vessels may depend on a vessel's size, elasticity, diffusion, and osmosis properties or the pulsations' frequency. For a hydraulic damper, modeling depends on many subsets such as the hydraulic chamber's length and valves' pressure.

1.5.2   DENDRITE GROWTH

Dendrite growth is one of the very familiar manifestation of a chaotic behavior and fractals that resembles the beautiful patterns of tree leaves, snowflake structure, or earth crust minerals and dendrite crystalline structures are examples of dendrite growth. Dendrite crystallization forms a natural fractal pattern that can be modeled by fractal mathematics. Sethian and Strain (1992) derived a numerical model for dendrite crystal growth. They started with metastable liquid under the freezing point and added a tiny seed of the solid phase of the material under study into the liquid. After the solid phase grows, it initiates a rapid formation of dendrites, stretching branches to colder regions of liquid. To model a crystal growth, one should consider parameters like crystalline anisotropy, the liquid's viscosity, time-dependent boundary conditions, time-dependent temperature, time-dependent viscosity of flow, surface tensions, and driving forces, and the kinematics of the particles involved.

Modeling dendrites through computation have been demystified the science behind different manifestations of dendrite in nature such as tree leaves and snowflakes and it was instrumental to the neuroscience research.

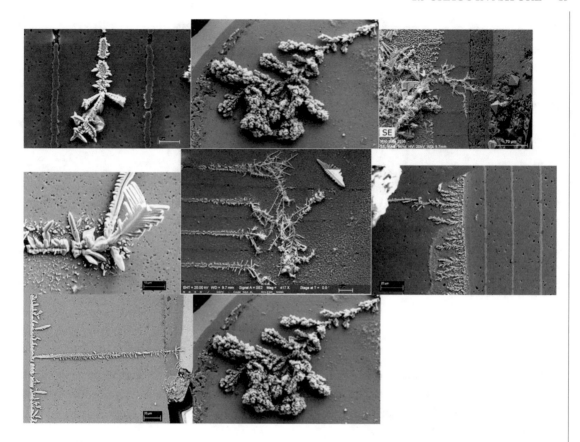

Figure 1.15: **Dendrite growth in ceramic capacitors.** Reproduced with permission from: Alexander Teverovsky NASA Space Technology: https://nepp.nasa.gov/files/24618/Dendrite%20growth%20 in%20BME%20and%20PME%20ceramic%20capacitors%20CARTS2013_n195.pdf.

Figure 1.16: Snowflake dendrites. Reproduced with permission from: Anita Stuckensmith.

1.5.3   CHEMICAL AGGREGATIONS AND SUPRAMOLECULES

The concept of fractals introduced by Mandelbrot was initially applied to the geometric description of chaotic iterative functions found its way to the field of dynamics. Initially described by Lorenz (Palmer, 2008) in the context of meteorology, fractal geometry applications range from fluid dynamics to biology and biochemistry.

Fractal topology is computational design approach that mimicks the self-assembly of proteins into tunable supramolecular fractal-like topologies in response to stimuli. Mimicking self-similarities of designed protein-based assemblies through fractal topology is a pathway to the essence of biomaterials' functionality (Hernández, et.al, 2019). Scientists also invented an imaging technique for taking a 3D view of the human genome and how DNA is packed inside cells. Research has shown that the DNA forms a structure called a "fractal globule," which is capable of holding vast amounts of material while remaining completely unknotted (Reuell, 2012).

Unraveling the mysteries of life's origin has been a long-standing effort. The evolution of the species and the mechanism of their behavioral development have been defined by natural selection, and the mechanism of biological systems are based on simple chemical and physical interactions. According to the Wyss Institute[1]: "One of the fascinating aspects of life is that all living organisms are formed through self-assembly, a fundamental biological design process by which an organized structure seemingly builds itself from a disordered collection of smaller parts. On a large scale, self-organizing behavior's powerful effects are seen when small gusts of wind join together to form a

---

[1]   https://wyss.harvard.edu.

tornado that can wreak havoc on infrastructure and natural resources in its path. On a much smaller scale, this same principle is seen when two strands of DNA zip up to form the double helix that encodes our genome. Alternatively, cells self assembles embryonic tissues that further develop into fully formed humans and animals."

It has been about a century since researchers began studying simple chemical aggregations as the building blocks of natural selection (Segre et. al. 2001). Recognition of macromolecules by another macromolecule, such as cells' interactions, self-assembly DNA duplication, antibodies, and antigen interactions, are mechanisms that keep organisms alive. Macromolecules in natures like synthetic polymers use covalent bonds and monomers to aggregate atoms and create a large assembly of molecules. The macroscopic behavior of polymers can be determined by their molecular structure and their intermolecular bonding. A branch of chemistry called supramolecular chemistry (Schmidt, 2012) specializes in noncovalent interactions. Noncovalent bonds can range from weak, such as dipole interactions to strong, such ionic or metal-ligand bonds. The weak and reversible forces—such as hydrogen bonds, hydrophobic forces, van der Waals forces, and metal-ligand coordination—are crucial to understanding biological processes, self-assembling systems, and molecular machinery. A minor interplay of external factors can affect the dipole and electromagnetic intermolecular interactions and, therefore, reverse the bonding. The macroscale features such as the phase of matter, viscosity, or self-assembly process are susceptible to temperature, the polarity of the medium, and concentration (Schmidt, 2012). As a result, a supramolecular chemist can take control of the parameters to initiate the self-assembly process.

1.5.4   METEOROLOGY

Weather forecasts are notorious for being unreliable. The scientific modeling for weather forecasts is a probabilistic modeling based on interrelationships of many physical and chemical components, such as gradient of pressure and temperature that lead to air currents, tides and gravity and changes in earth's axis of rotation, and also humidity and activities that habitats perform across the planet. Weather forecasting has a special place in developing chaos theory as Lorenz (Palmer, 2008)—a mathematician and meteorologist—discovered; it is a nonlinear behavior of weather conditions and its sensitivity to initial conditions (Buizza, 2008).

1.5.5   ECOLOGICAL COMPLEX SYSTEMS

Biological processes such as evolution, growth, or change in landscapes are spatio-temporal processes. Before fractal geometry was discovered, Euclidean geometry method was adopted in modeling structure which involves approximation. For example, modeling tree trunks with a cylinder or cone may be sufficient for a forester, but not for an ecologist whose research interest is in habitats living in the trunk or the truck skin structure fractal tool is more applicable (Kenkel and Walker,

1996). According to Kenkel and Walker, it is possible to estimate the fractal dimension of natural objects, such as landscapes and habitats, plant root systems, and the path trajectories of beetles, to develop models and test theories of landscape complexity which in turn predicts the spread of disturbance, abundance relationships in organisms, the movement of organisms and so forth.

1.6   SUMMARY

Understanding the behavior of a dynamic system involves generating a cognitive representation of a dynamic system. Various toolkits such as plots, contour maps, theoretical models, sculptures, optical imaging, mathematical toolkits, and fractal topology have been used for generating a mental model replica of a dynamic system and its constituents' interrelationships. Additional visualization tools have also been designed, e.g. simulations on physical phenomena and data visualization techniques have become valuable tools in many areas of physics.

Computer modeling has become an experimental component in modeling the iterated functions and solving complex differential equations. Whereas in older times, experts in the field had to rely solely on their imagination and their intellect, now mathematicians have an invaluable additional resource to investigate dynamics: previously, the computer and 2D designs and nowadays, 3D graphics designs with more live features such as merging digital rendering with the surroundings.

Scientific advancements have always been entangled with the developments of technology or other scientific or mathematical tools. Volumetric and holographic interactive renderings will soon become an essential technique for scientific communication as a new chapter in human–computer interaction technologies. But how can complex systems be understood using volumetric and holographic interactive renderings? A new era has arrived which opens a whole new vista for dynamitists. Unmasking the invisibles with the new era technology is the subject of thefollowing chapters.

CHAPTER 2

Spatial Abilities and Success in Sciences

The study of complex systems necessitates various cognitive and spatial skills. Unfortunately, the cognitive and learning challenges associated with complex systems are an understudied topic, and there is a need for further research in these areas. Preliminary research suggests significant learning challenges imposed by counter-intuitive components of complex systems. Understanding the behavior of a system involves generating a cognitive representation or, in other words, mental representation of the entities and their interrelationships.

This chapter will analyze dynamic thinking and the types of cognitive structures associated with various types of systems to identify a toolkit for unmasking complex systems.

2.1 EXPERT AND NOVICE

In the previous chapter, we categorized dynamic systems as deterministic, complicated, and chaotic. Dynamic systems can be categorized into different groups, and so it goes for the cognitive toolkits that are different for each group. Understanding a required dynamic system's behavior involves generating a cognitive representation of its constituents and their interactions. For instance, what comes to your mind first when you look at nature's colorful stones? Do you consider them spectacular antiques for decoration or a supply for a jewelry gemstone? What do you think would come to a geologist's mind? Perhaps, a geologist holds a different perspective, as a geologist is equipped with a cognitive toolkit that enables him/her to analyze the rock as a snapshot of a formation process. The image tells a story of transformations that have occurred through the ages. A geologist is equipped with a mental model of the motion and dynamicity of these layers that have been preserved since ancient times.

Figure 2.1: Jaspilite banded iron formation. Reproduced with permission from: James St. John Creative Common attribution 2.0, https://tinyurl.com/n277c88y.

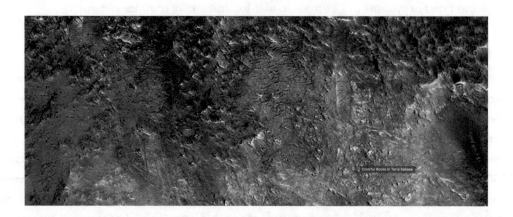

Figure 2.2: Colorful rocks in Terra Sabaea. Reproduced with permission from: NASA.

The shear zone traces give the geologist a hint about the motion of fractured regions that have been inactive for millions of years. Looking at the picture in detail, one may discover the hints, traces, and footprints and trace back the events that have shaped the rock. A geologist would put together the traces and hints to put up a storyline as million years ago, and these rocks were buried deep when they were warm and ductile, and different ductile layers were sliding past each other.

A mathematician may look at the picture and think about crafting a mathematical expression that could capture the flow geometry regarding rock's formation, or they may think about the fragmentation, age of the rock, or the corrosion process. A physicist may think about parameters such as viscosity, diffusion, melting and freezing point and pressures, and osmosis, which are due to the different responses of minerals to metamorphic conditions or changes in viscosities. A physicist may further think of how some minerals, like quartz or calcite, could flow at temperatures hundreds of degrees Celsius lower than other minerals.

The example above illustrates the details experts may consider in a snapshot of processes. It also illustrates that experts from different disciplines have different types of cognitive toolkits at their disposal. The cognitive toolkit that an expert uses determines his/her point of view. In an interdisciplinary team, often experts from various disciplines communicate using different toolkits. Therefore, interdisciplinary communications can benefit from having a cognitive translator. A cognitive translator can be an educator who is aware of interdisciplinary communication barriers and could create pathways to elucidate communication.

Subject matter experts have access to sizeable knowledge and cognitive skills with recognizable patterns and organization. Having access to these knowledge structures, in turn, affects how they notice the patterns of causal effect, which is not noticeable to novices. Based on their study context, experts can hold different points of view, observing the same phenomenon, and sketch a different mental visualization of the phenomenon's current and future behavior.

The same principle applies to other fields such as art and music. For example, countless musicians and poets have been inspired by nature and tuned their melodies and lyrics to reflect on its beauties, and a painter creates a work of art to express how they see nature. When a professional musician listens to a piece of music they may think about how the composer sets the lyrics in a particular way, with those particular sets of intervals making up the melody. A professional musician may ask the composer why they chose the exact chords for the harmony. Alternatively, what could be the possible story behind the lyrics? What does the song mean on the surface, and what is the subtext or subtexts?

How can the singer bring out the meaning they want to evoke by using different timbres and tone qualities and dynamics, or how does the songwriter add inflection to stress the essential words or notes? However, a nonprofessional musician may listen to the same piece and let the song trigger specific memories.

Similarly, when a physicist sees a detailed fractal set such as shown in Figure 2.3 and 2.4, they may find it a mysterious expression of mathematical iterations, which is numerical modeling of a system including some types of chaotic behavior.

Then physicists may also compare the pattern below with other manifestations of chaos which can be found everywhere, even in galaxies or marine creatures such as nautilus. In contrast, a fabric art designer or artist may look at the fractal set and find it a gorgeous design for artistic expression of inner inspirations.

Figure 2.3: **Partial view of Mandelbrot set.** Reproduced with permission from: Wolfgang Beyer, accessed at https://commons.wikimedia.org/wiki/File:Mandel_zoom_14_satellite_julia_island.jpg.

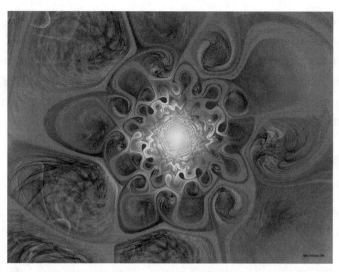

Figure 2.4: Fractal set. Reproduced with permission from: Tiffany Mcfarlane, https://www.trmdesignco.com/apophysis-fractals.html.

2.2   SPATIAL ABILITY SKILLS AND LEARNING ABOUT COMPLEX SYSTEMS

Comprehending the dynamicity of natural processes requires visuospatial thinking skills. *Spatial ability* is defined as the skill of solving problems in spatial forms (Carroll, 1993). Assessing spatial ability cannot be done in a holistic way as spatial ability is an amalgam of many factors and surveys that have been developed to gauge spatial ability factors independently. Various aspects of spatial ability are mental visualizations of a 3D object, rotating objects, spatial configurations of interacting systems, speed of scanning and comparing figures and symbols, and dynamicity of interacting system entities. Some other typical aspects of spatial abilities recognize noise and collect information from speedy and slow transitions (Hegarty, 2014).

Mathematical models, graphs, and formulations are usually compact and localized representations of cause-and-effect relationships among the events or detected patterns in the span of time and space. Representations are not only tools for demonstrating data structures but also operations to help visualize patterns and relations. The capacity of translating one representation to another or maneuvering among representations is a visuospatial skill that allows the learner to infer the equivalence interpretation of multiple representations such as real-time, diagrammatic, sentential, or math formulations.

Given the extended use of a graph, diagrams, operators, and mathematical representation of data in science and engineering, cognitive scientists (Hegarty, 2014) have conducted longitudinal studies and psychometric assessments to explore the role of spatial skills in scientific achievements. Many studies provided strong evidence that spatial ability is a predictor of success in sciences (Wai, Lubinski, and Benbow, 2009). The generalizability of the research outcomes on spatial ability can be achieved as the following psychometric reported results are applicable and in agreement with physics education research findings. The relevant research in physics education confirms the same results, which is the subject of discussion in the following chapter.

2.3   MACROSCOPIC AND MICROSCOPIC MENTAL VISUALIZATION

Hegarty (2014) emphasized the distinction between large-scale and small-scale spatial abilities. Large-scale tasks deal with occurrences such as navigating a route or visualizing the overall layout of a structure. Classical physics often deals with large-scale objects such as studying aerodynamics, architecture, civil engineering, planetary science, or working or designing experimental tools. Often, microscopic and macroscopic features are not mutually exclusive, like steam engines that contain combustion chambers. Within the combustion chamber cylinder, there is a mix of fuel and air ignited. The macroscopic features of combustion chambers involve the steam engine machinery and apparent features of the flame. However, the diffusion of the mixture of fuels and oxygen is microscopic.

Small-scale spatial skills are the ability to visualize the microstructure entities and their dissociations and associations. One with spatial skills in small-scale tasks does not always have good skills in large-scale tasks. Conversely, good spatial skills for larger tasks do not necessitate good skills in smaller-scale tasks (Hegarty et al., 2006; Hegarty and Waller, 2005).

2.4   VISUALIZING TIME SPAN, SLOW MOTION, OR A SPEEDY EVENT

According to Hagerty (2014), visibility and transparency of the scientific concepts can be obscured either when the speed of an event is too fast or too slow. Dynamic systems can be too fast and too slow and advancements in high-speed imaging allows capturing frame by frame the processes that are too fast for the human eye to discern.

2.5   MENTAL VISUALIZATION OF DYNAMIC PROCESSES

Visualization of sophisticated topics such as mechanisms of complex systems requires a combination of imagistic and analytical thinking. Clement (2009) defines the analytic thinking process as a mental animation of models created from static imagery. Inevitably, the learner could make incorrect mental animations when they make inferences about a dynamic system by putting together static imagery and speculating about the dynamic of spatial interactions (Hegarty et al., 2015; Janelle et al., 2014).

Contreras et al. (2003) distinguished between static and dynamic spatial abilities, which dynamic spatial abilities refer to the reasoning about a stimulus in motion rather than reasoning about a static diagram.

2.6   SPATIAL SOUND

Our brain deciphers the location of the objects around us not only by our eyes but also our ears. The sound waves bouncing back from objects arrive with path difference and different phases of waves that cause a sense of spatial sound. Considering the spatial sound effect in design will lead to more feeling of being realistic immersion (Ong, 2017).

2.7   TYPE OF SPATIAL SKILLS REQUIRED IN VARIOUS DISCIPLINES

Cognitive scientists identified different facets of spatial ability and noticed that various science domains required different aspects of spatial abilities. For example, in medicine, the ability to visualize anatomy cross-sections is significantly associated with the ability to test mental rotation that evaluates the ability to rotate mental representation (Caissie et al., 2009).

Studies (Hegarty, 2014) further showed a significant correlation between the flexibility of closures and performance in general chemistry and organic chemistry. The "flexibility of closures" is one of the items on the test of spatial visualization measuring the ability to recognize familiar patterns, objects that are hidden under distraction, and obscurity of surroundings. Correlations were mainly on the problems that required spatial representations of molecules, but they were not concerned with stoichiometry equations.

A correlation between "flexibility of closures" and students' success in geology (Orion, Ben-Chaim, and Kali,1997) is not far beyond expectation due to the demands for spatial reasoning such as inferring the internal composition of rocks and planets based on their outcrops or visualizing sliced view of various structures and evolution of metamorphic rocks.

Ample evidence supports that visualization plays a crucial role in physics problem-solving. Kozhevnikov et al. (2007) conducted studies in physics to examine the correlation between spatial visualization and solving 2D kinematics problems. The tasks required analyzing the motion of an object, collecting the data, representing the trends in various forms, and interpreting and extrapolating the kinematics graphs. Kozhevnikov and colleagues reported a significant relationship between spatial visualization ability and solving kinematics problems with multiple spatial representations and their findings confirmed by other studies (Duffy et al., 2020), which stated that students with high spatial abilities outperformed the low spatial abilities in physics problem-solving skills. However, the performance gap in physics problem solving between low and high spatial ability students vanished after students were instructed via well-equipped visualization techniques. Therefore, using technology for generating visualization companions improved the learning experience for students with low spatial abilities.

Physical problems are marked by dynamicity, and visualizing dynamicity can infer the integrations of different components called mental animation. Hegarty and Sims (1994) found correlations between spatial ability and mental animation of a working machine with all the interacting components.

2.8   COMPARING MENTAL VISUALIZATION OF CONCEPTS OVER DIFFERENT SPANS OF TIME AND SPACE

Visibility and transparency of the scientific concepts can be obscured either when an event's speed is too fast or too slow (Hegerty, 2014). Other studies have also confirmed that it is harder to comprehend the scientific concepts that occur over a long span of time. For example, in teaching scientific notation (powers of ten) in physics, the learner needs to comprehend the span of time and space and integrate it with fractions of time and space or massive to negligible weight particles. With AR, students can travel through ten factors and compare the time scales and connect those time scales to the phenomenon that occurs in each episode.

2.9   REPRESENTATION COMPETENCE

Numerous studies recognized students' lack of competence in navigating among multiple representation as one of the main reasons of students' problem-solving deficiencies. Multi-representational systems use various representational toolkits to distribute or compact information in a more manageable way. However, using multi-representational systems impinges upon how learners integrate information among representations (Ainsworth, 2006). The learner is expected to relate the displayed information in various formats and combine them into a sensible mental model. To understand the key features that influence students' difficulty in relating representations, Ainsworth distinguished the features of the representations in terms of their potential cognitive challenges. Ainsworth stressed that several factors are needed to be considered, such as an optimum combination of sensory channels of representation, modality of representation, dynamicity, and dimensionality of the representation.

2.10   SUMMARY

Cognitive studies provide a better understanding of the root of challenges facing learners in understanding and using spatial representations. Cognitive studies inform intervention designs of how to best tailor spatial thinking in science. Spatial reasoning is a central component of studying dynamic systems, and scientists have developed a large number of data visualization techniques to display trends and interrelationships of abstract data.

Our intervention designs are needed to be based on a better understanding of spatial ability and the key features that contribute to low and high spatial ability. In the following sections, you will find insights on how to design visualizations to either target low-spatial individuals' or help them use the designs more effectively or assist them in internalizing visualizations.

CHAPTER 3

Science Education Literature on Visualization

The premise of this chapter is to review science education literature on various aspects of unresolved challenges or inabilities in visualization. The aim is to build a stage for rigorous intervention designs. In the previous chapter, different facets of spatial reasoning were highlighted, and their relationship with teaching science and the dynamic concepts and processes were distinguished in terms of their spatial reasoning demand.

This chapter is a quick overview of guiding principles from a science education standpoint. Research in science education emphasizes evidence-based approaches, quality interventions, and assessment tools to evaluate the learning outcome. Studying the visualization challenges from two standpoints of science education research and psychometric unveils many commonalities between the outcomes obtained from two separate research paradigms, confirming the outcomes' credibility. Developing quality interventions necessitates several steps to identify the type of dynamic system and spatial reasoning skills. The next step is to implement evidence-based research, including science education and human-computer interaction principles. The final step is choosing adaptive technology to resolve the targeted challenges to unmask and reveal the invisible concepts and processes.

3.1   ACTIVE LEARNING AND VISUALIZATION

One of the most noted literature in science education is the premise of teaching by inquiry. The inquiry history goes back to John Dewey (1910), who proposed shifting in teaching objectives from knowledge accumulation to acquiring thinking skills and his ideas served as a cornerstone in developing an inquisitive teaching approach. In this way, studies to validate the inquiry approach were prompted and a large number of inquiry-oriented curricula were developed.

Further studies set out to determine the practical implementation of inquiry and essential aspects of teaching inquiry and further research provided abundant evidence to approve active learning environments' effectiveness (Freeman et al., 2014).

However, after a century, previous research of active learning vs. lecture-based had given its place to investigate how active learning instructional practices should be designed (Brewe and Sawtelle, 2018). While there can be flexibility toward approaching active learning, modeling instruction (MI) (Hestenes,1987) and problem-based Learning (PBL) (Goodnough, 2007) are two versions of recognized learning instructional practices in informal and formal science education. These two approaches have been shown to be efficacious and effective transformative learning methods used frequently and have a long history of scientific research to reinforce them. According to Brewe and Sawtelle (2018), the inquisitive essence of MI favors students in developing conceptual models and engaging in the practice of doing physics.

The theoretical and philosophical rationale underlying inquiry teaching is constructivism. MI and PBL approaches that emphasize the conceptual structure were the heart of EdTech solutions that later developed to create evidence-based learning environments.

3.2   ACTIVE LEARNING AND CONCEPTUAL STRUCTURE

One of the fundamental goals of active learning is constructing a quality conceptual structure using many types of invisible concepts in sciences. Factors that make conceptual learning challenging are often associated with a lack of spatial skills. Matloob Haghanikar (2012) conducted a study that modeled students' conceptual structure based on the parameters of instruction based on Lawson's categorization of concepts. Lawson et al. (2000) proposed categorizing scientific concepts based on their visibility. Lawson was a biologist and noticed a hierarchy of concepts associated with their level of difficulty. In this view, he categorized three types of biological concepts. The first type was descriptive concepts, including tangible and observable concepts such as mass, growth and development, heat, and light rays. The second type is invisible theoretical concepts that are deducted from cause and effects, analogies, or derivations from other theories such as genes that fall into this category. Lawson defined hypothetical concepts in the middle ground that are not usually observable, but they represent a visible pattern over time. Examples of these concepts would include natural selection, or evolution or other explanations of events that manifest themselves on a geological time scale. Lawson's classification was initially applied to the field of biology. To implement Lawson's scheme to the realm of physics, McBride et al. (2010) altered the definition of hypothetical concepts to include those that can be indirectly measured or observed, such as voltage, magnetic fields, and electric fields.

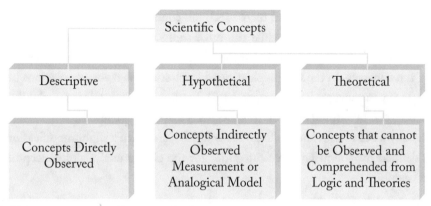

Figure 3.1: Concept categories. Courtesy: Author.

Lawson and Thomson (1988) reported a correlation between retention and the visibility of concepts. Furthermore, Lawson hypothesized it would be easier for novices to grasp descriptive concepts. This result has significant implications for instruction to start with descriptive concepts and follow hypothetical and theoretical concepts. In other words, in any field of study, or any platform, lecture-based or media platform, descriptive concepts precede the construction of hypothetical and theoretical concepts. In other words, instruction should introduce descriptive concepts before another type of concept.

Ausubel defined *meaningful understanding* as the quality of connecting new knowledge to what the learner already knows. Inspired by Ausubel's (1977) definition of meaningful understanding, Nieswandt and Bellomo (2009) expanded on Lawson's classification of concepts. They proposed not only categories of concepts but also types of connections among concepts contributing to their abstraction levels. They analyzed the extended written responses of 12th-grade biology students to questions. Nieswandt and Bellomo's analysis showed that difficulties transcend concept categorization and link the concepts, proving to be even more challenging.

Their findings have significant consequences for the assessments of students' abilities to reason scientifically because the evidence of understanding does not just involve demonstrations of the various types of knowledge but also connect the pieces and construct a mental animation of the pieces of knowledge. The quality of link making is based on how meaningfully students relate the concepts to each other with plausible explanations among the chains of cause and effects. In short, this occurred only when the students were able to explain what happened, why it happened, and how the causes related to the effect.

Nieswandt and Bellomo distinguished three types of links among concepts. A one-concept-level link refers to connections between two concepts from the same category (e.g., descriptive and descriptive). Cross-concept-level connections are connections between two different categories of concepts (descriptive, hypothetical) and multi-concept-level links occur when all three categories of concepts (descriptive, hypothetical, and theoretical) are connected. The least sophisticated concept link is a one-concept-category link, while the most sophisticated is the multi-concept-category link. Nieswandt and Bellomo postulated that meaningful answers must reflect multi-concept-level links among the concepts. In the least sophisticated level, students made connections between two same-level descriptive concepts (one-concept-level). Then, they examined students' abilities to make connections between two different levels of concepts (cross-concept-level).

Inspired by a Nieswandt and Bellomo study (2009), and with a slight modification, Haghanikar (2012) defined the following three categories for the cluster analysis of conceptual structure that students exhibited.

- **High-level-links** = Types of conceptual structures that consist of at least three meaningful links among higher level concepts such T or H, including one possible connection to D, the concept structure link for this groups includes structures such as; D-H-T, T-T-T, H-H-H, T-H-T, T-H-H, T-T-T-T, etc.

- **Middle-level-links** = Types of conceptual structures that consist of one meaningful link between H,T,D, excluding the case D-D. This group includes structures such as T-T, T-H and H-H or D-H.

- **Low-level-links** = Types of conceptual structure that includes only meaningful links between descriptive concepts.

Haghanikar researched students' conceptual structure and classified their responses in terms of the cluster analysis categories. The participants were preservice teachers who were taking a science course, and a total of 900 students around the U.S. participated in the study. The assessment was based on written response questions administered through their exam. The questions were designed to probe a certain level of conceptual structure and cognitive processing within each domain. Most of the students were able to connect descriptive concepts or some connections between theoretical and descriptive concepts. The occurrence of more sophisticated responses with multi-level links among concepts (theoretical-descriptive-hypothetical) was rare.

Sometimes students' responses involved lengthy explanations and hence contained a large number of concept links. To interpret the messy data and extract the valuable information, Haghnikar classified cases into six groups. Each group contained answers that were relatively similar in the types of concepts that the students used and how they linked the concepts. The groups were based on the following definitions.

- **Group A** = Answers that included two or more than two high-level links.

- **Group B** = Answers that included one high-level link in combination with middle-level links.

- **Group C** = Answers that included two or more middle-level links.

- **Group D** = Answers that included one middle-level link.

- **Group E** = Answers that included just low-level links.

- **Group F** = Answers with discrete concepts with no links.

Using the above classification scheme, Haghanikar obtained the relation between the quality of concept links were sought in relation to the degree to which the courses utilized active learning approaches.

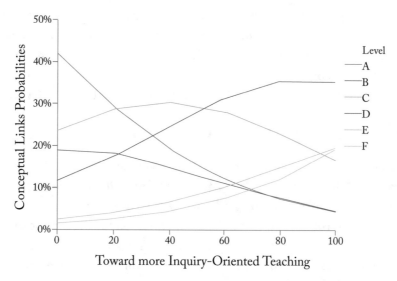

Figure 3.2: Concept level links vs. parameters of instruction. Courstey: Author dissertation.

Figure 3.2 has many stories to tell about the characteristics of instructions and students' meaningful reasoning. The color-coded graphs demonstrate the graphical representation of the probability distributions for individual probabilities based on cluster analysis multinomial logistic regression (Haghanikar, 2012). The curves "A," "B," and "C," represent higher-level and middle-level links. As we can see, all three curves show a decrease as we move toward classes that approached active learning. However, the lower-level links, which are "D," "E," and "F," show an increase as we move in the same direction.

The graph demonstrates that higher-order conceptual structures were less prevalent in courses that implemented active learning. With a hasty conclusion, this result may seem in favor of traditional teaching as students in traditional classes were more likely to exhibit higher-level concepts than students in reformed classes. However, from another standpoint, Haghanikar (2012) also analyzed students' reasoning levels while they applied theoretical concepts. Further analysis showed that student implications of using higher-order conceptual structures in traditional classes were associated with memorization and rote learning rather than thinking and reasoning.

On the other hand, for inquiry-based classes, much of the instruction involved observation and experimentation, and while students developed higher levels of cognitive reasoning and meaningful learning, they applied those learning skills to just visible concepts.

In sum, students in traditional classes exhibited higher-order conceptual structures, but their cognitive processing skills did not go far beyond memorization. Whereas in active learning classes, middle-level links were more prevalent, but those lower-level links were associated with higher levels of cognitive skills on the bright side.

The cluster analysis explained above agrees with Hegarty's (2014) assertations on lack of emphasis on spatial abilities. Hegarty noted that despite the importance of spatial thinking in scientific disciplines, promoting spatial thinking has not been stressed enough in our educational system. Lack of emphasis on spatial thinking hinders weak students from achievement in science and prevents educators from identifying and nurturing the most spatially capable students' talents.

3.3   SUBSTANCE-BASED ONTOLOGIES

Disfigured visualization and communication challenges between novice and expert have been a reoccurring problem in teaching and research. Slotta and Chi (2006) proposed that novice and expert reasoning under two different ontological categories. Processes and events belong to different ontological categories in comparison to ontologies of objects and matter. However, students tend to treat processes like substance-based concepts. Processes in nature are different from substances, and they are a set of indications that define an event. For example, heat, electric current, evolution, and light propagation are processes.

Abundant papers in sciences education research have reported students' difficulties with processes (Gupta et al., 2010). One of the difficulties in visualizing dynamic processes is adhering to constraining ontologies and replacing the dynamic attributions with static.

Gupta et al. argued that misinterpretations of ontological categories have severe implications for presenting the dynamic process and called for appropriate intervention. The scientific processes are dynamic; therefore, in addition to mental visualization of the 3D nature of non-matter processes such as magnetic fields, students need to create a 3D mental animation. We can find numerous well-known papers in science education research that reported students' difficulties with processes. For example, Wittman (2002) studied students' challenges in learning about mechanical waves and reported that students describe wave pulses as typified by an inappropriate interpretation of the wave pulse as a single, unified, pseudo-solid object.

3.4   CHANGE BLINDNESS IN VISUAL COGNITION

In visual cognition, change blindness refers to a perceptual phenomenon that happen when a change in a visual stimulus is introduced and it goes unnoticed by the observer. Recognizing information from the background noise is one of the facets of spatial ability that was discussed in the previous chapter. Similarly, physics educators landed on the same results when they compared the speed with which experts and novices detect physics diagrams' changes (Morphew et al., 2015). Researchers concluded that students might need a guiding schema of physics knowledge to direct their attention to relevant information. The same principle applies when interdisciplinary researchers communicate. A guiding schema is needed to navigate the researchers in a new field. Feil and Mestre (2010) compared the novices' and experts' responses to change. Experts and novices cue in on different aspects of the problem. While experts cue in on underlying principles, novices cue in on the superficial attributes.

3.5   EMBODIED COGNITION

Embodied cognition theory postulates that cognition is an interaction of the body and mind that takes place within the context of a specific environment (Herrera et al., 2018). Embodied cognition states how a body and its interactions with the environment affect human beings' cognitive activity (Soler et al., 2017). Therefore, instead of imagining an experience from the perspective of another, the viewer can see the experience from the embodied point of view of another. "Embodied cognition is a theory for understanding how our brains perceive as our bodies interact with the surroundings and therefore could give us useful clues about how to better design environments for learning" (Soler et al., 2017).

This perspective suggests designing not only for the mind but also for the body interactions as well. Previous designs included embodiment cognition with interactive graphic computer simulations. These types of kinesthetic activities improved learning (Lindgren et al., 2016).

Nowadays, implementing embodiment encourages movement or required feedback in addition to interactions similar to video game playing or interacting with robots.

3.6   REPRESENTATION COMPETENCE

Often, the learner fails to find connections between scientific models and real phenomena. According to Redish (1994), "Physics as a discipline requires learners to employ a variety of methods of understanding and to translate from one to the other—words, tables of numbers, graphs, equations, diagrams, maps. Physics requires the ability to use algebra and geometry and go from the specific to the general and back. "Therefore, changing representations and connecting the relationships among various representations has been reported troublesome for the learner (McDermott and Shaffer, 1982).

Students struggle to apply the relevant math in physics or see , and equations (McDermott and Shaffer, 1982).

3.7   CONNECTING MICROSCOPIC AND MACROSCOPIC FEATURES

According to cognitive science, there is a difficulty to visualize the microscopic world and how the microscopic phenomenon converges to macroscopic features (Wu and Shah, 2004). For example, in nanobiotechnology, a learner should find connections between electron microscopic images and the trends of statistical data and graphs. Science major students who are preparing for medical schools need deeper levels of interaction with microscopic entities.

Johnstone (1991) suggested a triangle including three domains in chemistry that are related to each other. These are the *macroscopic domain*, which is concrete and tangible; the *microscopic domain*, which includes the molecular, atomic, and kinetic parts; and *symbolic* or representational chemistry, including the use of symbols, equations, stoichiometry, and mathematics. Johnstone studies revealed that students faced challenges to maneuver among three levels of representations at the same time. Later, Reid and Yang (2002) proposed that interventions should be implemented to help visibility of macro-micro-symbol triangle.

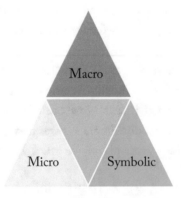

Figure 3.3: Micro–Macro–Symbolic triangle. Courtesy: Author.

3.8   SUMMARY

Science education literature has much to contribute to Human–Computer Interactions (HCI) and deserves a central role. As a result, it would be sensible to disclose a debate about the relevance of discipline-based science education research to HCI research. A summary of a checklist of design implications can be summarized. Instead of treating the checklist as a recipe, use it as an inspiring guideline for thinking and being creative. Multi-model designs include many dimensions, modes, and layers that need consideration in a well-balanced amalgam of various factors. The factors include, but are not limited to: enhancing spatial abilities, providing chances for active learning; interactivity, and opportunities for constructing knowledge: applying knowledge to new contexts, embodiment, exploration, data collection, data analysis, distributing information to avoid memory overload and sensory overload; and embedding social interactions, different representations, and multimodality.

It is worth noting here the importance of developing meaningful assessments for measuring the objectives' fulfillment. Science education research and cognitive science have provided deep insights and models for thinking and guiding principles for a design. Therefore, educational research's value is in the framework it provides to think about creative designs instead of a short recipe list.

CHAPTER 4

EdTech Solutions

Instructional technology is a practice of employing appropriate technology for education. The designing tactics may take many different forms based on the audience and the technology in use. To design an effective TechEd solution, instructional designers often rest upon the pedagogical principles and science of HCIs and cognitive theories. Sometimes designers get distracted by the aesthetics and exciting aspects of new technologies and overlook the research reports. Paying more attention to technological aspects rather than pedagogical principles will lead to a well-packaged learning asset with poor design quality (Margaryan et al., 2015).

During the past several decades, technological advancements have rapidly responded to the call for reforming traditional teaching, and EdTech solutions have emerged in various forms and capacities. Reflecting on the decades using EdTech solutions in sciences and revisiting EdTech resources collection can shed light on future designs.

This chapter is a quick overview of the past designs with the technology advancements that occurred over the last two decades. Admittedly, this brief review cannot do justice to include all interventions that have been evolved through several decades. Nevertheless, it could serve as an open discussion for a meaningful design for the features needed to be considered. Reading this chapter, bear in mind the communication barriers that have been discussed so far and conclude what type of barriers have been addressed.

4.1 SCOPE OF EDTECH DEVELOPMENTS IN STEM

A quick review of the spectrum of existing EdTech solutions is beneficial as a scaffold for future advancements and developments. Due to the prevalent need at the time of COVID, the American Association of Physics teachers created a shortlist of EdTech solutions based on the collections that were inventoried in the comPADRE. Most of these educational technologies have been tested and evaluated, and their effectiveness and shortcomings have been reported.

A broad spectrum of EdTech innovations that addressed core communication barriers included sensor-based devices, simulated games, microscopic videos, highspeed imaging analysis techniques, microcomputer labs, virtual labs, and interactive books. The following is a quick introduction to each of these technologies.

4.1.1   MICROCOMPUTER-BASED LABS

Microcomputer labs (MBL) emerged with the advancement of manufacturing microprocessors with integrated sensors. The technology of attaching a sensor to a computer became available (Thornton, 1987) about several decades ago and right away was introduced to the realm of science education. The sensors were a significant addition that enabled precise data collection and data analysis and modeling. Using an MBL configuration allowed data to be collected and analyzed in real time. Data collection occurs via sensors that can be connected to a computer or iPad using built-in or separate interfaces. The variety of sensors such as force, heat, sound, EKG, pH, frequency, and speed has made real-time experimentation possible in all disciplines (Bernhard, 2003). Nowadays, MBL labs have been commercialized using various computer platforms to perform real-time experiments and not simulated ones. The cognitive aspects of the MBL computer-assisted lab help students simultaneously see different representations and compare and contrast the real-time event with its equivalent graphic or tabular representation. The images below show the PASCO MBL labs on classical mechanics.

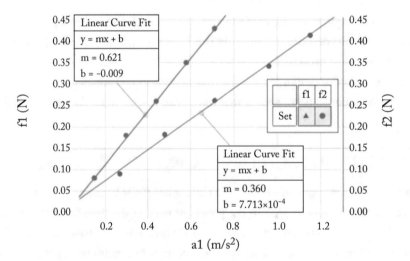

Cart 1 Mass = 0.357 kg and Cart 2 Mass = 0.607 kg

Figure 4.1: PASCO labs. Based on and courtesy: PASCO Scientific.

The capability of MBL to represent data along with real-time experiment and transform it immediately into a graph is one of the greatest advantages of using MBLs (Brasell, 1987). Students can simultaneously analyze a graph being plotted at the same time they are conducting the experiment.

4.1.2   CAPSTONE PROJECTS ON ABSTRACT CONCEPTS WITH MBL

Physics instructors often arrange for creative demonstrations to enhance visualization using traditional experimentation or MBL sensors. For example, the dipole moment is a purely mathematical concept, and students tend to learn how to manipulate the mathematical representations. In the previous chapter, we categorized concepts based on their visibility. Students usually have difficulty visualizing the rotation and polarization of dipole moment in an electric field. I used carbon fibers in a hydrophobic solution and placed the solution in a 5,000 v/m electric field. Students can observe the shift in the laser beam orientation passing through the solution by turning on and off the electric field, using a PASCO sensor that measured the light intensity. In this way, students measured the net effect of dipole polarization and the rotation angle and made a better sense of purely mathematical representation. However, the problem with this type of project is that they are not scalable as they often need bulky instrumentation. In the following chapters, we discuss how to make a scalable demonstration using augmented reality apps.

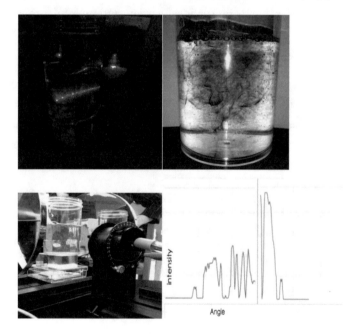

Figure 4.2: Visualizing polarization of dipole moment. Courtesy: Author

4.2   SENSOR-BASED SIMULATIONS

Sensor-based simulation (SBS) kits display illustrations with simulations using the data collected by the sensors. SBS kits and MBL experiments are different because SBS kits simulate more instruction aspects than the experiment itself.

For example, Chao et al. (2016) developed a teaching kit for kinetic molecular theory and the gas laws using an SBS. The SBS kit they called gas frame collects data from three input sensors. The gas frame is a simulated replica of an actual gas chamber, and the simulation is coded to interpret the sensors' input as if the actual chamber had been exposed to those changes. The microprocessor transforms the input information collected from the sensors and generates a simulation exhibiting the gas's microscopic behavior as if it was really enclosed in the gas frame. The intervention was intended to connect the microscopic features to the macroscopic. If we use an actual chamber for the experiment, we would not be able to display the microscopic features, limited to algorithmic and quantitively solving equations rather than qualitative reasoning to solve problems (Sanger and Greenbowe, 2000). The macroscopic parameters are volume, temperature, the force on the syringe, and gas pressure. The pressure is used to detect gas compression or decompression change that changes the number of gas particles in the syringe, and a force sensor is used to indicate the force on a piston on the gas chamber that was simulated in the virtual world. Furthermore, students would be able to perform various experiments on the ideal gas law and molecular theory.

4.3   SCIENTIFIC GAMES

Gamification uses game elements and game design techniques in a non-game context (Tolentino and Roledo, 2019). Gamification can improve students' engagement and be an excellent resource to spice up outside of the classroom activities. In the current dynamic of innovative designs, gamification has found its unique place as a pedagogical sound tool for active learning (Adriana, 2015).

Example of Scientific Games

The theory of special relativity is one of the most fascinating but least accessible topics in astrophysics as it is often explained purely theoretically. "Special relativity" implications often conflict with ordinary senses (Scherr et al., 2001). The theory compares observers' perspectives in relative inertial frames of reference, speeding close to light's speed compared to each other. Lorentz equations can measure the appearance of the objects observed in the fast-moving frame approaching the speed of light. Using a mathematical logic of special relativity is straightforward. However, visualizing the counterintuition that it defines is a matter of concern.

Even though teaching special relativity makes a substantial break from Galilean relativity, both in structure and its concepts,it still heavily relies on an understanding of inertial frames. Some of the difficulties learners encounter in learning "special relativity" associated with the p-prims students often bring to learning a new context, such as robust beliefs in absolute motion and time. Students tend to believe that the time of distant events is determined by the time order in which the observer perceives signals from the events, and in this way of thinking, events are simultaneous just when the observers receive the signals at the same time.

In what follows, many physicists attempted to create simulated experiences of special relativity. For example, Wegener et al. (2012) designed a physics game software package that engaged the user with aspects of time, space, and light noticeably different from our familiar physical reality.

4.3.1   INTERACTIVE SIMULATIONS

The use of simulation in teaching physics has become prevalent during the last two decades. Physlets and PhET simulations are two more common research-based materials recognized by physics educators. These materials provide a new, exciting, and effective way to deliver interactive curricular material to students studying physics. The effectiveness of these materials has been evaluated and tested and evaluated. Physlets[2] are specifically designed for physics; while the majority of PhET[3] simulations (Sims) is designed for physics teaching in the majority, it has recently developed simulations for chemistry, biology, math, and other sciences as well.

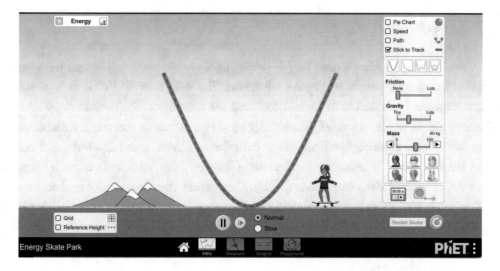

Figure 4.3: Energy Skate Park. Reproduced with permission from: PhET Simulations, PhET Interactive Simulations to distribute this content under the terms of the Creative Commons Attribution License available at http://creativecommons.org/licenses/by/4.0/.

---

[2]   https://www.compadre.org/physlets/.
[3]   https://phet.colorado.edu.

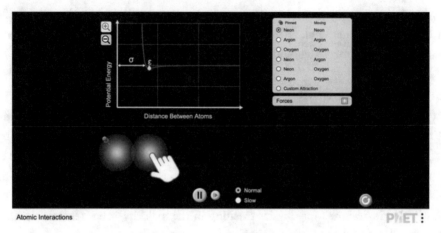

Figure 4.4: Phet simulations: Atomic Interactions. PhET Interactive Simulations to distribute this content under the terms of the Creative Commons Attribution License available at http://creative-commons.org/licenses/by/4.0/.

Visual Quantum Mechanics (VQM) project (Zollman et al., 2002) is another open-source simulation based on research in physics education. The design of VQM aims to simplify the quantum to make it accessible to high-school and introductory college students by eliminating mathematics in its presentation. VQM provides hands-on activities, computer visualizations, and written worksheets are integrated into an activity-based environment. Zollman and colleagues utilized the Learning Cycle instructional model (Karplus, 1974). Students begin by exploring a modern device—Light Emitting Diode (LED). They realize that they cannot explain the device's operation using any existing metal models and, therefore, must construct new models to explain their observations. Subsequently, they apply these mental models in different situations. Physlet, VQM, and PhET are open sources and designed based on the research report on students' misconceptions and challenges.

Labster[4] is a collection of simulation laboratories in various disciplines of science. The software runs via computer simulation or VR headset. The virtual labs include experimentations that can be found in real science labs enriched with conceptual visualizations.

[4] https://www.labster.com.

Figure 4.5: Labster Ideal Gas Lab. Reproduced with permission from: Labster

4.3.2 VIDEO MICROSCOPE

To help life-science students find the relevance of physics to their biology courses, Moore et al. (2014) developed a set of laboratories and hands-on activities using video microscopes. Moore et al. (2014) selected biological contexts with physics experiments to relate physics concepts to biology and adopted the particle-tracking velocimetry method. For example, the freshly cut onion skin was placed under the microscope and the intracellular motion of tiny organelles inside living cells was recorded. The motion of the vesicles was tracked, analyzed, and modeled. There was additional room to explore the relation between a motion to energy and explore biological implications of the physical measurements and estimate the ATP hydrolysis rate involved in the process (Redish, 2014).

4.3.3 BIO KITS

With the invention of biosensors, interactions with bio chambers through computers and IPads have become possible. Bio entities respond to triggers such as electric and magnetic fields and heat. In this view, a new generation of remote labs was generated. In this view, Lee et al. (2015) invented an interactive microbiological living cell through touch-screen experiences. Users can interact with Euglena gracilis—single-celled phototactic microorganisms that are stimulated by a microscope setup equipped with a projector and a touch-screen display. This setup allows users to draw patterns on the screen and interact with these organisms.

4.3.4   ARDUINO

Arduino is an electronic board combined with sensors, expansion boards, and a software development environment (Organtini, 2018). The chip is programmable using the Integrated Development Environment (IDE) software, freely available from the Arduino website, and can be transferred with a USB. The function of Arduino depends on the code user write on their personal computer, and it compiles it and then transfers the program to the Arduino memory.

Sensors for temperature, humidity, pressure, piezoelectric, light, sound, acceleration, magnetic field, and currents can be cheaper alternatives for various physics experiments, ranging from calorimetry to electromagnetism, from thermodynamics to the physics of waves (Organtini, 2018). Arduinos originally were not designed for physics education but have been considered increasingly since their pedagogical value was recognized.

4.4   SUMMARY

Until 60 years ago, there was just one way to teach the sciences: through the single-mode transfer of information which was rigid and conventional labs with almost similar sets across universities. However, research in science education, cognitive science, and technology advancements have radically changed the teaching and learning platform with much room for creativity. Many scientists have understood the importance of learning theories and have started implementing them in their courses. There are always a group of people who resist change, and many institutions and funding agencies have been so far slow to respond. However, with the COVID-19 pandemic outpouring, many organizations reconsidered EdTech solutions because of the urgency to transition to online.

The online education industry was surging even before the pandemic with low-quality online labs, with many educational institutes facing lack of instrumentation and space. In addition, there will be a new demand for teaching more sophisticated concepts as we are on the verge of industrial transformation with education not keeping the pace with industrial advancements (Haghanikar, 2019).

CHAPTER 5

Emerging Technologies: A Twist on EdTech Solutions

The advent of new of technologies has constantly revolutionized education and research.

Educational technology is a rapidly evolving field, and the advancement of technology has been always applied to the field of education with various levels of success. During the last few years, media have been reshaped rapidly into using volumetric mixed reality platforms which includes virtual reality (VR), augmented reality (AR), and HoloLens. To this end, the global market of AR/VR games is expected to grow, and many still frame AR/VR reserved for gaming. Nevertheless, the use cases of AR/VR go far beyond gaming. Different organizations such as healthcare, the auto industry, museums, and many more have started leveraging on the benefits of mixed reality (Muikku and Kalli, 2017) technologies.

Over and above, the unprecedented transformation of computer interaction technologies has also provided a new stage which can replace the contemporary means of scientific communication and data visualization among experts and between expert and novice.

Mixed-reality technologies have also redefined contemporary data presentation platforms and allowed users to interact with data from different perspectives. Data visualization, which is the essence of scientific communication, is entering a new era. The advancements of artificial intelligence have found its way to immersive and interactive data visualization and a way to solve challenging scientific problems. A boost in visualization has provided easier clarifications on data analysis and more effective inferences in scientific communication.

Stepping into a new revolutionary era of HCIs can also bring out the best of curriculum design.

This chapter reviews the hardware, software and key strategic parameters of various VR/AR devices with highlighting typical examples that have been developed.

5.1 VIRTUAL REALITY

VR is a medium by which developers can share their imaginations with users in volumetric and immersive fashion. Synthetic renderings of VR replace the current view of the user with the simulated view or full immersive 360° virtual view. One of the great advantages of VR is to enable users to completely immerse into sceneries landscapes, places, and positions which are not easily attainable.

Does it make sense to rush out and buy the most expensive VR head that you can afford? Definitely not necessarily. More expensive VR headsets require installation and a motion tracker. That may not be a problem if you specified a room, but a bulky instrument could be a heavy load if you need to carry it up and down many flights of stairs, want to take it on an airplane, or store it in a cramped lab storage. Therefore, if you just need a VR, for a journey to touristy places, or to visit the museums, travel on a spaceship, or shop for a house, or immerse yourself under the sea, etc., you may not need a VR headset with a motion tracker and a simple headset would suffice. But if you want to use VR for an interactive project such as traveling on a spaceship and interacting with objects, you would need a space with controllers and motion trackers.

Some of the newer VR headsets have improved motion sickness and latency issues and that is an important parameter for consideration regarding the age of the user.

5.1.1   VR HEADSETS

Google Cardboards are the simplest headsets developed by Google. The Google Cardboard can be set up simply by folding the predesigned Cardboard, downloading Cardboard app from Google Play, and inserting the cellphone into the viewer's enclosure in the cardboard. The Cardboard app splits the scenery streaming from the cellphone in two identical screens and if the lenses are located at the optimal place from the screens the user's brain infers a depth and dimension that is added to the 2D sceneries. The mechanism of the brain perceiving depth is associated with watching the same images or scenes from two slightly different views. The Google Cardboard app splits the mobile screen in two and streams the identical content from each screen. In this way, our brain infers a depth by looking at identical scenes through the lenses of Google Cardboard from two slightly different perspectives. Google Cardboard was revolutionary in the way that it was affordable and empowered the innovative VR developers, however, with the advent of advanced headsets, Google Cardboard lost its market except for the VR marketing goodie bags.

The successor of Google Cardboard was Google Daydream with some controlling features and nicer fabric which also has been discontinued. Google Cardboard, Daydream and its rival Samsung Gear all were mobile-based VR headsets which did not catch consumers' interest. One setback for mobile-based VR headsets is restricted possibly to use their cellphone at the same time when their phone was placed in the headset.

In the last decade, many major technology companies have started developing AR/VR technology. The attention of these primary technology companies to AR/VR has improved the efficiency and quality of the design and construction of VR headsets. Generally speaking, there are two categories of VR headsets:

- stand-alone VR: all necessary components to provide a VR experience built in the headset; and

- headset with computer/VR: the headset by itself does not have all the necessary components to provide a VR experience; the headset functions as a display device and is connected to a primary device such as a computer.

One of the major VR headset competitors is Oculus. Oculus offered three headset models namely, Oculus Rift S, Oculus Quest, and Oculus Go. Oculus Rift VR headsets are computer based and require touch-based controllers, as well as a tracking system that allow users to interact with the 3D environment. Oculus Rift-S, which consists of a fully immersive headset with two controllers, requires a high-performance computer. The Oculus Rift-S utilizes the computer for higher-end graphics and performance placing the user in a fully immersive experience. The other version is Oculus Quest, which consists of a fully immersive headset with two controllers and a stand-alone device not requiring a computer.

One of the Oculus' distinctive competitors is HTC Vive that uses room-scale tracking technology and motion-tracking controllers so that a user can move in 3D and interact with the virtual landscape.

There are many other evolving models and trends on VR headsets, however the premium features of VR headsets can be summarized as a higher number of pixels, the latency, the degrees of freedom, and the field of view. Having lower latency is crucial for head-mounted VR sets and if the system is slow to respond to the head movement it will cause motion sickness for the user.

Another feature is degrees of freedom. In the real world, three degrees of freedom refers to the motion of rigid body in the 3D coordinate system.

5.1.2   EDUCATIONAL USE OF 360° CONTENT

Spherical videos or 360° videos are omnidirectional videos recorded by panoramic cameras. The terms 360° and VR are often used interchangeably, but they are fundamentally different. With 360° videos a user can roam around and immerse themselves in omnidirectional videos. The content of 360° video isn't a virtual world but is an actual video from the real world. For example, using Navigation Cameras, NASA's Perseverance Mars[5] rover stitched together individual images from Mars after they were sent back to Earth. The resulting video can make a VR journey to Mars possible.

5.1.3   SIMULATED VIEW VR

Simulated view VR provides users with the perception of being in a different reality by displaying a digital 3D immersive environment that is a fictional reality through 3D graphics. This category of VR is used often for educational or training purposes and relies on high-resolution renderings that immerse users in a computer-generated alternative world.

---

[5] See https://www.youtube.com/watch?v=IX6LEAqUx-E: Curiosity Mars Rover's view atop Mont Mercou (360 View).

5.1.4   INTERACTIVE SIMULATED VIEW WITH CONTROLLERS

The VR headset can be equipped with additional controllers to support interactions of the user with a virtual world. Leveraging on technological advancements there is a variety of hardware that provide tools for more meaningful interactions. Fahmi et al. (2020) distinguished among three types of interactive controllers called VIVE controllers, Leap Motion controllers, and Senso Gloves controllers. The VIVE controller uses buttons to trigger instructions on the system. In addition, the VIVE controller can provide feedback in the form of vibration to the user. Leap Motion controllers use an infrared sensor detector that tracks hand motions connected by a USB cable to the computer. When the user's hand is in the field of view of the detector, the hand gestures are sensed by an infrared camera. Unlike VIVE controllers, Leap Motion controllers make no contact with the user's hand and as a result no feedback can be felt by the user's hand. The third type of controller introduced by Fahmi et al., is the Senso Glove, which is a wireless glove that can detect the user's hand motions. The Senso Glove works with a Bluetooth network that can send information to computer. In addition, the haptic feedback generates vibrations that can be sensed by the user. The vibrator motor is mounted on the Senso Glove fingertips, backs, and wrists.

Fahmi and colleagues compared VIVE, Leap Motion, and Senso Glove controllers on the VR content that was developed for teaching anatomy. Results showed that the VIVE controller was superior compared to the Leap Motion controller and Senso Glove in terms of usability, ease of learning, movement suitability, and display suitability. However, the Senso Glove was more favorable in terms of haptic feedback satisfaction.

There are many examples of VR contents with controllers. For instance, NASA developed VR simulations that give users an immersive experience of walking in space station[6] or an experience of interacting with an asteroid, exploring life on other planets, lia fe under the sea, swimming with sharks, or traveling into the human body.

5.1.5   INTERACTIVE SIMULATED VIEW WITH HAPTIC

Simulation VR with haptic offers another dimension to the virtual experiences, which is the realization of touching sensations. By simulating other senses such as touch, smell, or simulating an environment with spatial sound, a computer can act as a gatekeeper to a new world that was not accessible before. VR creates a simulated view that triggers our senses, and the first generations of VR headsets started with stimulating our sight and sense of hearing.

The immersion experience is more realistic with headsets equipped with haptic feedback accessories to activate a third human sense, touch, and haptic gloves or jackets.

There are many examples of VR haptics for education. For instance, Seo et al. (2019) designed an interactive VR experience to train people to use fire extinguishers. The VR experience

---

[6]   https://www.youtube.com/watch?v=WX9_Nwwew7s.

concerns a simulated view of fire hazard using a head-mounted display (HMD), including kines-thetic experiences using a pneumatic muscle and vibrotactile transducer. The VR fire extinguisher is designed to train users to interact with the fire extinguisher when the fire breaks out. A user could experience an illusion of a fire scene. Besides, when the handle of the fire extinguisher is squeezed to release the extinguishing agent, the haptic device generates both vibrotactile and airflow tactile feedback signals, providing the same experience as that obtained while using a real fire extinguisher.

5.2 HOLOLENS IN SCIENTIFIC COMMUNICATION

The science and art of sculpting holograms using laser coherent beams and light interference patterns has been around for several decades. The age of light-sculpting holograms boomed along with the emergence of laser technology in the 1960s. However, the holograms discussed in this book are digital holograms emerged in 2015 when Microsoft announced mixed reality platform which can be viewed by HoloLens. In compare to VR headsets, HoloLens is a transparent headset. The transparent displays allow the user to see the 3D holograms that are blended with the real surroundings. There is a lim-itation in the field of view of HoloLens, therefore, Ong (2017) suggested a creative design that could compensate for this shortage. Ong suggested increasing the immersive features and adding visual cues.

Figure 5.1: Microsoft HoloLens. Reproduced with permission from: Ramadhanakbr is licensed under CC BY-SA 4.0, https://commons.wikimedia.org/wiki/File:Ramahololens.jpg.

5.3 zSPACE IN SCIENTIFIC COMMUNICATION

The pieces of hardware required for zSpace[7] technology are a zSpace laptop, a stylus pen, and zSpace light eyeglass. Every zSpace system has tracking built into the display. The built-in sensors monitor the zSpace stylus pen and the eyeglasses. As the users change orientation to browse the ob-

---

[7] zSpace.com.

ject, the software adjusts the perspective to construct a whole and high-definition 3D view. zSpace comes in two forms, laptop and all-in-one computer.

Looking through the zSpace eyeglass, digital objects seem to be floating out of the laptop screen, and users can interact with the objects with the stylus pen. The position and orientation of the 3D objects can be explored as they are real objects, and the user can adjust their view to obtain a realistic and comfortable viewing experience. In this setting, the user can get practical experience in assembling the parts of a device.

With the zSpace setup, the user can get practical experience in assembling the parts of a device. At the same time, zSpace can be used for visualization of abstract interactions. An example can be a 3D visualization of spherical coordinates in math, representation of Lorentz force in physics, replicating Zeeman or Stern Gerlach experiments for students to recognize the classical and quantum mechanics spin, assembling pieces, and parts of lasers or pipes in engineering.

zSpace Laptop

zSpace All-in-One Computer

The zSpace Lab

Anatomy Lab

Figure 5.2: The zSpace 3D Lab. Reproduced with permission from: zSpace.

5.4   AR

The most compelling platforms of AR devices for education are Google AR[8] and image recognition AR, Vuforia, which we are planning to examine and utilize. The titles below are a detailed description of each product line.

---

[8]   https://www.youtube.com/watch?v=ZtJzS33S-rk.

5.4.1   GOOGLE EXPEDITION: AUGMENTED REALITY CLASSROOM DEMONSTRATIONS

Google has developed a series of rendered science lessons using the AR app that scan the physical classroom and add virtual objects to the classroom. Google also has developed several Google AR science kits that have been used for science demonstrations in classes. Among the demos were 3D visualizations of DNA, visiting astronauts, and volcanos. The demos were 3D digital renderings that merged into real time, allowing teachers to guide students through collections, 3D objects, or experiments. Students could look through their cellphones or iPads and interact with the 3D rendering that has merged into their world.

In the same way, we are going to implement the Google Expedition AR innovation to perform lecture demonstrations for undergraduate science classes. A simple illustration is a demonstration of magnetic fields around the wires carrying electric currents. Students usually have difficulty visualizing the direction of magnetic fields as a result of currents. With an AR app, an imaginary augmented wire will appear as a real object in front of the class, and students could interact with the current and its magnetic field via their tablets or cellphones. Another use case for engineering education can be seen inside the combustion chambers. Students could point their cellphones or iPads toward chambers and watch the series of interrelated processes that occur insider the chamber.

5.4.2   APPS OF AR ON CELLPHONES OR TABLETS

For this series of products, we use image recognition-based AR. The ARcore is a platform for developing AR apps for Android devices, and ARkit is the platform of Apple's version. We use image recognition AR technology to overlay 3D information on a specific target. For example, students could hover their phones around an image to prompt computer-generated 3D visual objects. Therefore, students can acquire more information about the image and view it from different angles. The various use cases include adding a new dimension to books and tutorials, exploring and comparing the molecular processes, or finding more information about an unseen process such as observing the inside of the combustion chamber. One of the use cases can be OpenStax and open sources. Many students every year depend on OpenStax resources, which are unembellished downloadable PDF documents. However, with using AR, their experience can be enhanced by viewing microstructure features or 3D visualization of objects.

To avoid memory overload, the display of concurrent scenes serves as an extended memory to help with comparing different instances. Closing the gender gap is another area to evaluate the effectiveness of AR in education.

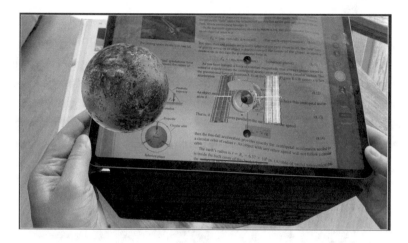

Figure 5.3: Adding dimension to textbooks. Courtesy: Author.

There are several degrees of freedom for the user to interact with the digital AR assets such as zooming in and out, rotating the objects, connecting the elements, assembling, navigating between the scenes, and prompting new assets.

The AR is a spatial type of a media that allows designs, which are floating in a space, and constitutes embodiment. According to the theory of embodiment cognition, sensory-rich experiences, being present in the virtual world, and building interconnections between the self and environment influence cognitive skills. For example, according to Moreau (2012) sensory motor experiences can shape cognitive processing skills.

Mixed reality technologies turn the physical environment into a digital playground that allows users to interact with virtual objects and investigate their relationships with the environment (He et al., 2018). Users can actually be in the imaginary environment, participate with all their senses and interact with actual surrounding and have an immersive experience rather than interacting via screen. It is also possible to enrich the virtual scene with contextual data, animation, or other types of digital asset.

5.5   DESIGN TOOLS

The first step to design is to draft a storyboard and create 3D graphical designing of the imaginary scene. PowerPoint and Keynote are good starting points for animations and annotations (Pell, 2017). For sketching 3D graphic designs there are several options such Cinema 4D, Blender, Maya, and 3D Studio for Sculpting and modeling. There are several tools for VR sketching such as Sketch or Quill[9] which is a tool for VR animation.

[9]   https://quill.fb.com.

5.5.1   UNITY[10]

Unity is a game engine platform for game design and creating VR/AR or HoloLens renderings. It also permits C programming and Java scripting as well as Python programming which may need Python for unity or another extension. Unity is user friendly as it is equipped with a built-in data studio and a pre-coded collection of Prefabs. Prefabs are digital packages that are coded for certain types of operations. Pref-fabs can serve as game objects to be saved in the unity working project for reuse. Prefab rendering can then be shared or purchased without having to be configured again.

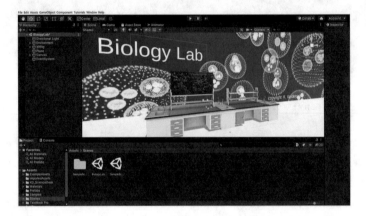

Figure 5.4: Unity editor. Reproduced with permission from: Karen Vanderpool-Haerle.

5.6   SUMMARY

The practical implementation of emerging technology can be achieved by foreseeing affordances of each technology and the scope of their application.

It is like when you decide to purchase a camera the most important aspect of any camera is its aperture and portability. A scope's aperture determines both its light-gathering ability, brightness, and resolution. Likewise, when you choose an emerging technology, the most important aspect is the type of immersion it provides and its field of view and latency. Various models provide different levels of immersion and interactivity.

Before choosing and developing an intervention, the designer must determine what's important to accomplish. How experienced are the users? How much are they prepared to spend? How much room is needed for storing the equipment and how much room is needed to perform a project? What is the scope of the problem and which feature of technology provides the best solution? Generally, feasibility studies precede the technological development and that may include conducting market survey, exploring what's on the market, and choosing a technology that will resolve the need of the users.

---

[10] Unity.com.

CHAPTER 6

Curriculum Design and Emerging Technologies

During the past decades, digital visualization and simulations were widespread tools to represent data visualization and communicate the complex concepts in physics and other sciences. However, emerging technologies have been fundamentally transforming the nature of HCI, offering new allowances that were not possible before. The new allowances can revolutionize the educational platforms. Recent advancements in digital holography, VR and AR, and immersive data visualization have unlocked the possibility of interacting with the third dimension. The recent advancements in VR/AR developments hold numerous potentials to enhance the visualization in sciences. The scope and significance of this chapter is on reviewing typical challenges in different disciplines through the lens of reasoning skills, spatial abilities, and cognitive processing skills, and suggest a typical design to address those challenges. For the interest of generalizability, typical challenges of different scientific fields are reviewed with an example of a suggested design.

6.1 AR PLATFORM CLASSICAL MECHANICS

As we discussed in Chapter 3, science education literature suggests that students often struggle to apply the relevant math in physics or see the connections between different representations such as interpreting a motion in real time, graphs, and equations (McDermott and Shaffer, 1992). Mobile AR or HoloLens offer a combination of several scenes' overlays on the original target. There are many ways to use AR for classical mechanics. One way is to use motion trackers and Arduinos.[11] A motion detector collects data which fed to an Arduinos electronic board. Arduino can be coded to present 3D graphs or other characteristics of the motion. In this way, users can make easier connections between different representations. In addition to provding better visualization, the set up suggested here is cheap, light, accessible, and can be used remotely as well.

6.2 BREAKTHROUGH DESIGN FOR PHYSICS

Quantum mechanics (QM) and special relativity usually deal with gedanken/thought experiments which is reasoning and analyzing a hypothetical scenario. Many great thinkers adopted thought experiments to explore impossible situations to predict their implications and outcomes. In this way, gedenaken experiments played an essential role for the inaccessibility of experiment topics such

---

[11] Arduino.cc.

as the initial development of the quantum theory and special relativity. Although, thought experiments are imaginary and based on abstract reasoning, they can be considered a precondition for advancements in experimentations. Nowadays, scientists often run thought experiments or proxy experiments when the real experimentations are impossible, like a twin paradox in special relativity or the Schrodinger cat in quantum mechanics.

In topics such as quantum mechanics or special relativity, our common understating of length scale is distorted. For example, in quantum mechanics length scale is defined in terms of matter waves or de Broglie wavelength which is not imaginable based on our everyday experiences. In special relativity, the concept of length is linked with time and relativistic speed. For the speeds near to speed of light, it is impossible to conduct the experiments other than powerful colliders and accelerators. Experimentation in quantum mechanics require well-equipped research labs. Replicating these labs for teaching purposes involves insurmountable challenges and restricting factors. As a result of these limiting factors most of the focus has been given to mathematical aspects or gedanken/thought experiments. Numerous studies in physics education have reported that often quantum mechanics courses have de-emphasized conceptual understanding (Zhu and Singh, 2012). Students might be able to master the Schroedinger equation for a variety of potential wells and yet struggle when it comes a conceptual understanding of basic concepts.

Immersive data visualizations have unlocked the possibility of interacting with the third dimension.

6.3   USING AR OR ZSPACE IN TEACHING STERN GERLACH EXPERIMENT

With new affordances of emerging technologies, the upgrades of hands-on activities are imminent. An AR rendering offers newer allowances to provide immediately critical intuition that will position its user into the quantum universe.

To demonstrate the potential of QM AR asset, we selected a sequence of Stern Gerlach (GS) experiments. Students often face many challenges in learning of Stern Gerlach experiment (Zhu and Singh, 2009). Stern Gerlach is conceptually simple and demonstrates basic ideas of QM especially fundamental ideas for starting quantum computing. In addition, it can be a supply for AU generating electrons with spin up down.

The Stern Gerlach AR rendering will incorporate variety of Stern Gerlach experimental settings and outcome results to verify the quantum theory that has been developed to predict those results.

In its simplest form, the AR replication of apparatus consists of an oven that produces random neutral atoms guided toward a field of an inhomogeneous magnetic field, and a detector screen as depicted in the figure below. Stern and Gerlach used a neutral beam of silver atoms to avoid the deflection of charged particles affected by the magnetic field. Classical mechanics predicts a continuum of distribution on the screen. However, the experiment shows that the beam splits into two. One beam is deflected upward and one downward as a result of the magnetic field gradient. The deflection of the beam in the Stern Gerlach experiment is thus a measure of the intrinsic spin Sz along the z-axis, which is the orientation of the gradient of magnetic field.

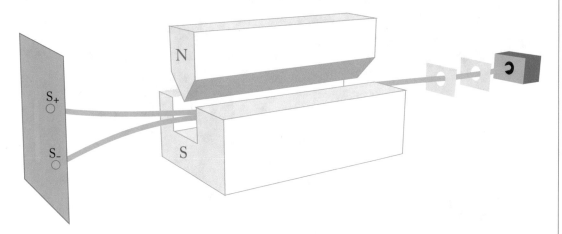

Figure 6.1:Simple Stern Gerlach experiment.

The AR replication of the SG experiment is a 3D arrangement of an oven with controllable oven temperature and a magnet with adjustable gradient of a magnetic field and a detector that depicts the distribution of the deflected electrons. Additional magnets with adjustable gradients of a magnetic field and adjustable orientation can be added to the experiment. Therefore, a series of experiments can be performed on other orientations of magnets along the x and y direction can be performed. This arrangement makes it possible for students to examine whether the operator Sz and Sx, or Sz and Sy, commute which provides an experimental alternative for a purely mathematical approach to quantum mechanics commutation relations.

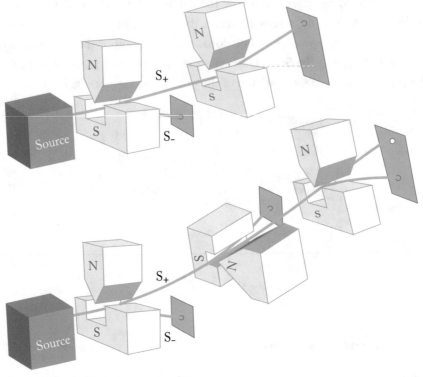

Figure 6.2: **Array of Stern Gerlach experiments.**

6.4   VR PLATFORM ASTRONOMY

The James Webb Space Telescope Virtual Experience designed a VR experience called the Web-bVR game exoplanet discovery module with additional Spanish-language options available for free download.

The WebbVR[12] is an interactive VR educational game, based on real data from NASA that immerse the user in a voyage to explore solar system planets and moons, galaxies, black holes, planet formation, stellar lifecycles—and now, extrasolar planetary systems.

The VR game is designed to teach users about the three astronomical procedures of discovering exoplanets orbiting distant stars. Exoplanets are often not observable, but their shape, size, color, atmosphere or density can be obtained with astronomical analysis. However, with WebbVR, developers put the finding into artistic design of surface appearance for each exoplanet. The planet and distance from the host star of the system are also embedded in the VR game.

---

12   https://webbtelescope.org/webb-science/the-observatory.

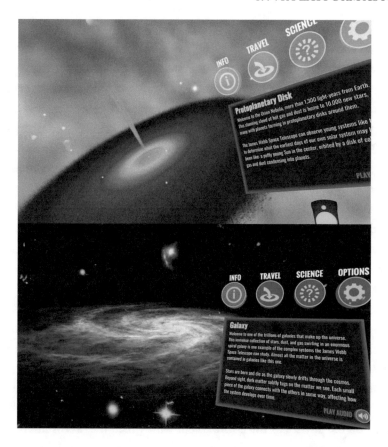

Figure 6.3: VR game for astronomy. Reproduced with permission from: WebbVR Space Telescope Science Institute.

Figure 6.4: VR game for astronomy. Reproduced with permission from: WebbVR Space Telescope Science Institute.

6.4.1   VR PLATFORM SPECIAL RELATIVITY

Virtual space is an alternative when the real-time experience of a phenomena is impossible. While it would be rather simple to gain mastery in mathematics using special relativity equations, it would not be easy to conceptually comprehend experiences that contradict common experience and accepting a model of time and space that is strange and unfamiliar (Mermin, 2005).

Wegener et al. (2012) developed a simulation experience of special relativity, called real rime relativity (RTR). RTR simulates the visual effects of the finite speed of light and special relativity when traveling at near light speed. RTR is a game-like experience where the user controls his/ her speed, direction of motion, and direction of view around a world built of clocks, planets, and abstract shapes. Within RTR, the Doppler and headlight effects can be toggled to avoid obscuring other effects and the speed of light may be set as infinite to allow the user to become accustomed to navigation controls and the virtual world. Otherwise, the user is immersed in an authentic VR where they can experience and experiment with visual, space, and time effects while traveling at near light speed.

For instance, the surrounding environment can respond to the leaner in a natural way, and one can experience traveling to the quantum level, experiencing the fast speed or threshold frequencies interacting with an inaccessible environment. Imagine virtually building an environment which immerses users in a situation that mimics high-speed motion and lets the users experience the physics of such an environment. We already have traditional simulations that have provided virtual experiences of abstract environments (Wegener et al., 2012). In this way, the surrounding's response is based on the physics of the environment with which the user is interacting.

6.5   AR/VR PLATFORM WORKFORCE TRAINING

Many industries have switched to emerging technology for workforce training tasks. The advancements that VR/AR has provided more cost-effective means. Industries can save time and space and increase capacity and eliminate the safety concerns.

6.5.1   VR PLATFORM REMOTE EXPERTS

One of the challenges facing the workforce training industry is access to external experts. To meet this challenge Elvezio et al. (2017) invented a remote subject-matter expert VR station to assist a local user. Elvezio and colleagues used a 3D referencing approach and allowed the remote expert to create and manipulate virtual replicas of physical objects in the local environment. The expert could guide the novice to assemble or fix the occluded parts. The alternative way, which is using video training, has restrictions when 3D spatial referencing is required. In contrast, AR/ VR allows 3D spatial referencing that assists local users to view live instructions directly overlaid on their environment.

Many industries have turned to cellphone AR for remote training manuals. The remote collaboration and knowledge-sharing platform allow remote field technicians and experts to connect and collaborate on maintenance, repair, operations, and inspections using the unique features of AR. To learn how to use specific instruments, the user points the cellphone toward the targeted instrument, and the scene becomes augmented and enriched with text and cues and the expert on the line. The platform connects the remote experts to novices, but it is also more accessible for the user to read the directions on the instrument rather than going back and forth reading from the manual and looking at the task.

Figure 6.5: Cellphone AR in workforce training. Reproduced with permission from: RE'FLEKT Inc.

6.5.2 VR/AR REPLACE BENCHES

Due to the industrial transformation, there will be a gap as workforce training is not keeping pace with industrial advancements. Training and on-boarding of new technicians and engineers require expensive benches. VR headsets can be used to replicate the benches and serve as an advanced organizer[13] (Ausubel, 1977) of the instrument-specific information with which technicians need to familiarize themselves.

Safety issues are another limiting parameter that hinders the experimentation. Among many advantages of advanced augmented labs is the possibility of performing dangerous experiments

---

[13] An Advance Organizer is a tool that instructors present to students before a lecture to structure the information they are about to learn.

without safety concerns. Engineers could practice experiments such as assembling pipes or observing inside a combustion chamber by aiming an iPad toward a combustion chamber.

Many industries have implemented VR/AR or HoloLens experiences in their training benches.

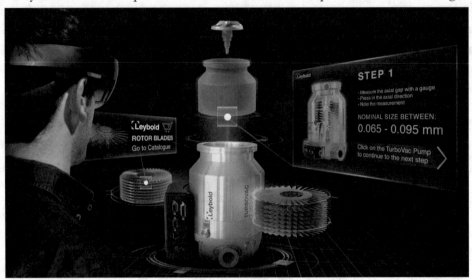

Figure 6.6: HoloLens workforce training. Reproduced with permission from: RE'FLEKT Inc.

Figure 6.7: Augmented Reality for industrial training. Reproduced with permission from: Creative Commons Attribution-Share Alike 4.0, https://commons.wikimedia.org/wiki/File:Entrenamiento-industrial-Fyware.jpg.

6.6   SUMMARY

Immersive visualization has unlocked the possibility of interacting with the third dimension; however, people tend to think about the third dimension when it comes to implementing applications. To name a few, with using interdisciplinary AR apps, students could see the abstract physics concepts, interacting with electron microscope imaging, replicating physics experiments, assembling laser parts, rule the biological realm, and affect their everyday life and medical students will gain a more profound vision about the physics of the biological phenomenon.

AR is a powerful tool for helping students connect the pieces of knowledge as they point their cellphone toward a specific target; illustrations appear as 3D animated frames and overlay on the target as they are prompted.

We speculate that the seamless overlaying of the 3D images prevents working memory overload and helps the learner compare and connect the frames. Hou and Wang (2013) used older versions of AR and argued that AR could also be used to close the gender gap by addressing the working memory problem.

AR helps retain the working memory for the novice assemblers who experience considerable working memory overload (Hou and Wang, 2013). AR's unique feature allows overlaying scenes and extended view and degree of freedom in distributing the data. Distributing the data serves as extended memory to help with working memory overload while comparing different instances.

CHAPTER 7

Breakthroughs in Scientific Communication

So far, we have discussed various toolkits that scientists invented to unmask the hidden features of dynamic systems. We differentiated two categories of dynamic systems that can be represented either with a differential equation or iterative functions. Fractal topology appeared to be one of the computational approaches that sheds light on the hidden patterns of chaos. Our current mathematical toolkits are broader now than when fractals were discovered about four decades ago, and it keeps growing. The field of mathematics is dynamic by its nature, and new ways of statistical modeling and computational techniques will be invented driven by the needs of complex systems such as biological models. We also discussed learning strategies that persuade visualization and promote system thinking. We differentiated dynamic thinking from static thinking as the accuracy of mental models, which depends on adopting an appropriate mathematical and computational approach and ontological categorization. However, all that has been accomplished so far to model the complex systems is just tipping the iceberg compared to what is needed to be accomplished. What follows in this chapter highlights a new paradigm in modeling complex systems and scientific communication, leveraging the unique affordances of emerging technologies.

7.1 CYBER LEARNING VS. E-LEARNING DEVELOPMENTS

Cyberlearning is a term defined by the National Science Foundation, and it represents using a genre of technology for implementing research-based principles to facilitate learning.

A new ecosystem of learning practices is evolving along with technological advancement and a tremendous potential for cyberlearning is rising. Past research informing a learning platform should include activities that embed interactions and experimentations. So far, e-learning focused on converting knowledge to a digital format that shifted the learning platform to more globally networked, connected, and remote. New opportunities arose for lifelong learning with a more connected learning network, such as micro-credentials or Citizen of Science. Micro-credentials provided pathways for individuals to develop their skillsets, and Citizen of Science were initiatives such as Zooniverse in which the public could participate and follow their research interests and collaborate with a scientist to interpret the data. With computational advancement, cyberlearning thrives from simulation and sensor-based experiments to more inquiry zone-adapted types of experiments. According to CIRCL (n.d.): "cyberlearning is the use of new technology to create effective new learning experiences that were never possible or practical before." The unique features of emerging technologies have allowed more flexibility in implementing research-based methodologies that were not possible before. Harnessing the power of computation, the highlighted features below are the affordances that were not attainable before.

7.1.1   3D USER INTERFACES

Contemporary communication tools which have been commonly used are illustrations, mathematical models, 3D model science sculptures, computer programming designs, simulations, and sensor-based experimentations. Complex systems are dynamic, and the 3D model sculptures that are frozen in time and static cannot be useful tools for exploring the interactions and dynamicity of changes in various dimensions.

There is a difference in how the brain processes the 3D mental models created by simulations and AR/VR. For traditional simulations, 3D is bounded to the screen, and the user should move the object around to collect information from different perspectives and compile the various aspects to internalize a 3D mental model. However, with AR, the 3D is augmented to the real world. Therefore, the user's field of view is extended in comparison to the screen and more flexible interactions are possible for the user and degrees of freedom for comparing and contrasting mental visualizations. From that place, the mental processing step of surveying and compiling different perspectives is eliminated with AR designs. It is also worth considering that interpreting the third dimension naturally comes from phase differences of rays that emit from a 3D scene, and the brain interprets it as the third dimension. With AR/VR, the process of seeing 3D matches the natural way of the brain, while with simulations, the process of inferring needs compiling's sceneries. Accordingly, there will be less cognitive demand and more room for further interactions.

7.1.2   LECTURE DEMONSTRATIONS

Now that Google AR can bring 3D digital objects in living spaces, the whole classroom management can be rearranged. Instead of looking at a static 2D image, students could see a 3D object in dynamic motion. Looking through scene viewer which can be an iPad or cellphone, learners could interact with the 3D representations or participate in an exploratory and interactive lecture demonstration. Interactive Lecture Demonstrations (ILDS) is one of the widely accepted methods that has emerged from a constructivist view of learning. Numerous research have shown the effectiveness of lecture demonstration methods in teaching sciences as students with ILDS intervention outperformed the students in traditional classrooms. Sokolof and Thornton (1997) defined eight steps for lecture demonstration as described below.

1. Present an experiment or a web-based demo.

2. Ask students to record individual predictions.

3. Have the class engage in small group discussions with nearest neighbors.

4. Ask each student to record the final prediction on handout sheet (which will be collected).

5. Elicit predictions and reasoning from students.

6. Carry out the demonstration with data analysis.

7. Ask a few students to describe the result. Then discuss results in the context of the demonstration. Ask students to fill out "results sheet" (which they keep).

8. Discuss analogous physical situations with different "surface" features. (That is, a different physical situation that is based on the same concept.)

Despite many pedagogical advantages, using the ILDS approach has been hindered due to innumerable challenges, such as difficulty in carrying the experiments to the classroom, parallel scheduling, lack of resources and manpower to maintain the experiments and to manage the inventory. In many universities, often classes are not situated in one location and distributed over different buildings. Oftentimes, lab technicians have many responsibilities and there are shortages regarding lab technicians for putting experiments together, space limitations for storage and maintenance. All these problems can be resolved by developing a digitally rendered platform of various demos. Using this technology students could point their cellphone devices toward a specific target and interact with demo. To name a few, consider illustration magnetic field around a wire, Gauss law, flux, electromagnetic waves, angular momentum, and spin and change of polarization with temperature.

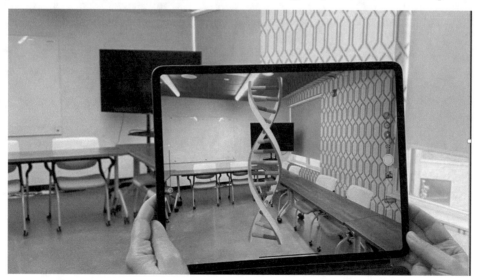

Figure 7.1: AR for Lecture demos. Courtesy: Author.

7.1.3   REMOTE LABS

The physics online courses often suffer from major setbacks associated with the restrictions of remote labs. The quality of science in remote labs is usually lower in comparison to the same on-campus science labs science educators have developed simulations, digital sensors, and video tracking techniques to export the labs and activities. At the end, due to the 2D restrictions of the screen, sensory memory overload (Kirschner et al., 2010), and lack of versatility for different learning styles the quality of the online science labs is lagging behind. A remedy to this situation is a type of more advanced digital asset that could provide multi-modality and tangibility. With developing the AR-VR platform, students explore the experiment using their hands, interacting with the surroundings or looking at the various perspectives by turning their heads around.

Safety issues are another limiting parameter that hinders the experimentation. Among many advantages of advanced augmented labs is the possibility of performing dangerous experimentation without safety concerns. Engineers could practice experiments such as assembling pipes in a safe environment.

7.1.4   EMBODIMENT COGNITION

The unique features of AR/VRs make these new technologies good candidates for a wider range of embodied interactions and kinesthetic integrations. For example, many VR war games allow for a full range of body movements, recent AR apps allow dancing with avatars mimicking the users' motions, a tilt brush allows drawing in space in three dimension, and functionality of the Microsoft HoloLens allows flexible options for embodied interactions.

Figure 7.2: Interactive VR game. Courtesy: author

7.2   SPATIAL REASONING AND EMERGING TECHNOLOGIES

In Chapter 2 we discussed the high correlation of different facets of spatial abilities with success in STEM field. Below specifically we focus on focus on specific aspects of spatial ability and explore how can they be improved by using these new technologies.

7.2.1   DYNAMIC SYSTEMS

Substance-based ontologies (Gupta et al., 2010), and misinterpretations of ontological categories that were discussed in Chapter 2, have implications for process-based concepts. Therefore, researchers who studied substance-based ontologies have called for effective intervention that introduces process-based concepts. The scientific processes are dynamic; therefore, in addition to mental visualization of the three-dimensional orientations, a mental animation of system's components are required. We can find numerous well-known papers in sciences education research that reported students' difficulties with processes (Wittman, 2002).

By merging and combining the digital scene and real-time scene in AR, learners can be guided to enhance their ontological categorization and articulate the infrastructure mechanism responsible for observable changes in the system. This articulation is called "mechanistic reasoning," which is vital to teaching by inquiry (Russ et al., 2008), and mechanistic reasoning is the heart of inquiry-oriented teaching strategies.

7.2.2   CHANGING REPRESENTATIONS

With augmented reality, we can combine several scenes overlays on the original target with the potential of navigating back and forth seamlessly. Furthermore, with implementing AR, merging the digital and physical world into one become possible and users can navigate concurrently in different representations. For instance, one can interact with the graphs and real-time event at the same time in 3D environment and explore the relationships.

7.2.3   CONNECTING THE MICRO TO MACRO

There are many advantages of using AR assets over simple animations. For example, AR can be like an offline microscope with an ability to zoom in and out seamlessly. Students are more accustomed to the real-world and with AR both real and digital world merge together, therefore, students could instantly build the connections between tangible and abstract world. For example, in order to relate the change of temperature on the light triggered bio-reactions one can generate AR 3D electron microscope imagery of low and high temperate of biochemical reactions. Given this view, the microscopic models become accessible to the user and become part of the student's real world.

7.2.4   ESCAPING SENSES

In Chapter 2, we talked about developing lessons with an emphasis on visualizing time span, slow or speedy events. The extended field of view in AR allows flexibility in putting different frames over time and comparing and contrasting to find the connections.

As a simple illustration, in teaching scientific notation (powers of ten) in physics, the learner needs to comprehend the span of time and compare the spectrum of masses from massive to tiny particles. With AR, students can travel through ten factors and compare the time and size scales and connect those scales to the phenomenon that occurs in each episode.

There are variety of data analysis software such as ImageJ, video analysis or Global Hands-On Universe, cutting-edge image analysis tools, and resources for statistical modeling of biological migrations, random walk, or comprehending the astronomical trends.

Figure 7.3: High-speed photography, colorful liquid. Reproduced with permission from: Alex Koloskov, https://www.photigy.com/school/.

Unexpected events can happen when the time duration span is too fast. For example, for short laser pulses, self-phase modulation, and nonlinear optics occur in high frequencies of a laser beam. The nonlinear phenomenon occurs when a laser beam propagating in a medium interacts with the medium and imposes a phase modulation on itself and it one of bizarre effects, which escapes common experience. According to Shen and Yang (2016): "The physical explanation of the phenomenon lies in the fact that the strong field of a laser beam is capable of inducing an appreciable intensity-dependent refractive index change in the medium. The medium then reacts back and inflicts a phase change on the incoming wave, resulting in self-phase modulation (SPM). Since a laser beam has a finite cross section, and hence a transverse intensity profile, SPM on the beam should have a transverse spatial dependence, equivalent to a distortion of the wave front. Consequently, the beam will appear to have self-diffracted."

7.2.5   MULTIMODALITY

According to Kress (2009), the multimodality theory examines how individuals process information through many modalities such as visual, audio, gestural, and spatial modes.

The individual differences which are of such practical importance to education vary in their ability to process information presented in various modalities. The essence of mixed realities is multi modalities, embodiment, immersions and interactions.

7.2.6   CHANGE BLINDNESS

Lack of fluency to extract the appropriate information from pictorial graphs and other representations in learning sciences or reading science papers can be remediated using augmented reality. By adding visual cues to the diagrams, we can provide an auxiliary navigating system for a user. Recently, Hyundai has exploited AR tech and visual cues to help users with maintenance. Similarly, students or non-experts could follow visual cues to detect and recognize the relevant information the same way as the experts do. The preference of using AR technology over past developments of assets with cues is an extended field of view of the playground and more degrees of freedom. Users can construct connections among various frames, namely, connections between virtual frames or connections between virtual and real-time frames or merging digital assets with the surrounding environments.

7.3   RESHAPING RESEARCH

7.3.1   DATA MODELING

One of the achievements of emerging tech in the realm of research is 3D data modeling that can lead to more accurate modeling. Schaad et al. (2017) created a 3D visualization from blood vessels down to the capillary level. Schaad and colleagues focused on the vascular network of the murine hind limb in 3D, and a multi-scale imaging approach was followed. It worth reading the various stages of imaging techniques they examined in their paper (Schaad et al., 2017)

7.3.2   RESEARCH REPORTS AND PAPERS

The challenge of data accumulations has compounded researchers at many levels. Scientific papers are not equipped with sufficient data visualization tools to effectively communicate research work and access the research environment. Manuscripts are hard to follow within the constraints of page limit and communication tool shortcomings. Sometimes it is even impossible to comprehend a scientific paper, especially when the reader is an outsider to the discipline. Scientific articles usually carry a specialized language and too much jargon, compressed into page limits, and inaccessible to the nonexpert. The researcher's thought patterns, and structure can be illustrated using AR guiding cues, and the reader can be guided to follow the thought patterns of the researcher. Another dimension can be added to research journals that overlay research experiment results on the top of narration.

7.3.3   VISIBLE THINKING WITH VR PAINTING WITH GOOGLE TILT BRUSH

Tilt Brush[14] is a Google painting VR application that allows users to paint with VR controllers in 3D. To use Tilt Brush, users need to wear headsets and create 3D sculptures[15] in the space. After sketching, the sculptures can be printed with a 3D printer. While Tilt Brush is developed for artistic application, there is a place for it in education. For example, for visibility of thinking, Tilt Brush can serve as a tool for evaluating students' mental models, spatial reasoning, or spatial abilities.

Until now, our emphasis was on designers to create VR assets for users. Now let us look from the user's point of view and use their 3D creations to evaluate their imagination, which can be an excellent assessment tool. Tilt Brush allows drawing 3D images that can be stored and saved. Tilt Brush allows the 3D artistic drawing and allows communicating thoughts and imagination in the physics realm, such as sketching a 3D electric field. Therefore, we are going to achieve a higher level of transparency in expressing our thoughts, which could open a new chapter in case study qualitative research and assessment.

"Informed by learning science, cyberlearning is the use of new technology to create effective new learning experiences that were never possible or practical before" CIRCL (n.d.).

7.4   SUMMARY

This book is a glimpse into the future of EdTech designs for overcoming barriers of visualization in modeling complex systems. The intention was not to catch up with the latest technology trends but instead focus on the compelling features of emerging technologies that can enhance visualizations compared to the previously adopted solutions. Implementing AR/VR in education is an exploratory endeavor, and its advantages and pitfalls are increasingly being reported. Thus far, the previous products have provided a foundation for future developments, but they failed to resolve many remaining challenges that are still barriers to system thinking. The future adoption of AR/VR technologies may seem uncertain; however, the previous research in science education and cognitive science strongly suggests the promising future of emerging tech.

[14] https://www.youtube.com/watch?v=TckqNdrdbgk.

[15] 3D sculpting in VR with Tilt brush, see https://www.youtube.com/watch?v=TckqNdrdbgk.

There is no magical technology that can resolve the barriers of visualization. At its best, for each case, the magic rests within the evidence-based designs guided principles grounded in research and the choice of technology suitable for targeted goals.

I hope this book has inspired you to stir up your educational insights for creating your breakthrough designs, exploiting technology's power with a well-calculated design.

References

Adriana, G. (2015). *Gamification as Teaching Pedagogical Practice in the Teaching and Learning Process in the Theme of Social Inclusion.* Universidade Tecnológica Federal do Paraná. Londrina. 42

Atcheson., B. (2007). *Schlieren-Based Flow Imaging.* The University of British Columbia. 14

Ainsworth, S. (2006). DeFT: A conceptual framework for learning with multiple representations. *Learning and Instruction*, 16(6), 183–198. DOI: 10.1016/j.learninstruc.2006.03.001. 26

Anadol, R. (2020). Art in the age of machine intelligence, YouTube. 5, 6

Ausubel, D. P. (1977). The facilitation of meaningful verbal learning in the classroom. *Educational Psychologist*, 12, 162–178. DOI: 10.1080/00461527709529171. 31, 63

Bandura, A. (1986). *Social Foundations of Thought and Action.* Prentice-Hall.

Berg, I. (2020). The Magnetic Pendulum. Beltoforion.de., https://beltoforion.de/en/magnetic_pendulum/index.php. 10

Bernhard J. (2003). Physics learning and microcomputer based laboratory (MBL) learning effects of using mbl as a technological and as a cognitive tool. In: Psillos, D., Kariotoglou, P., Tselfes, V., Hatzikraniotis, E., Fassoulopoulos, G., and Kallery, M. (Eds.) *Science Education Research in the Knowledge-Based Society.* Springer, Dordrecht. DOI: 10.1007/978-94-017-0165-5_34. 40

Bransford, J., Brown, A. L., and Cocking, R. R. (2000). *How People Learn: Brain, Mind, Experience, and School.* National Academy Press. 1

Brasell, H. (1987). The effect of real-time laboratory graphing on learning graphic representation of distance and velocity, *Journal of Research in Science Teaching*, 24(2), 385–395. DOI: 10.1002/tea.3660240409. 40

Bratianu, C. (2007). Thinking patterns and knowledge dynamics. [accessed Apr 04, 2021]. de Barcelona, Spain. https://tinyurl.com/wab2cnxn. 3

Brewe, E. and Sawtelle, V. (2018). Modeling instruction for university physics: Examining the theory in practice. *European Journal Physics*, 39. 29, 30

Buchbinder, S. B., Alt, P. M., Eskow, K., Forbes, W., Hester, E., Struck, M., and Taylor, D. (2005). Creating learning prisms with an interdisciplinary case study workshop. *Innovative Higher Education*, 29, 257–274. DOI: 10.1007/s10755-005-2861-x. 1

Buizza, R. (2008). The value of probabilistic prediction. Atmospheric Science Letters, 9: 36-42. DOI: 10.1002/asl.170. 17

Caissie, A. F., Vigneau, F., and Bors, D. A. (2009). What does the mental rotation test measure? An analysis of item difficulty and item characteristics. *The Open Psychology Journal*, 94-102. DOI: 10.2174/1874350100902010094. 25

Carroll, J. (1993). *Human Cognitive Abilities: A Survey of Factor-analytic Studies*. Cambridge University Press. 23

Chao, J., Chiu, J. L., DeJaegher, C. J., and Pan, E. A. (2016). Sensor-augmented virtual labs: using physical interactions with science simulations to promote understanding of gas behavior. *Journal of Science Education and Technology*, 25(1). DOI: 10.1007/s10956-015-9574-4. 42

Chi, M., Slotta, J., and Deleeuw, N. (1994). From things to processes: A theory of conceptual change for learning science concepts. *Learning and Instruction*, 4(1), 27-43. DOI: 10.1016/0959-4752(94)90017-5.

CIRCL (n.d.). Retrieved 4/4/2021 from https://circlcenter.org/new2cl/. 67, 74

Clement, J. (2009). *Creative Model Construction in Scientists and Students. The role of Imagery, Analogy and Mental Simulations*. Springer. 24

Contreras, M. J., Colom, R., Hernandez, J. M., and Santacreu, J. (2003). Is static spatial performance distinguishable from dynamic spatial performance? A latent-variable analysis. *Journal of General Psychology*, 130. DOI: 10.1080/00221300309601159. 25

Danielsson, A. T., Engström, S., Norström, P., and Andersson, K. (2020). The making of contemporary physicists: figured worlds in the university quantum mechanics classroom. *Research in Science Education*. DOI: 10.1007/s11165-019-09914-9.

Devaney, R. L. (2018). An Introduction to Chaotic Dynamical Systems: Second Edition. DOI: 10.4324/9780429502309. 9

Dewey, J. (1910). Science as subject-matter and as method. *Science*, 121–127. Retrieved April 5, 2021, from http://www.jstor.org/stable/1634781. 29

Duffy, G., Sorby, S., and Bowe, B. (2020). An investigation of the role of spatial ability in representing and solving word problems among engineering students. *Journal of Engineering Education*, 109(3), 424-442). DOI: 10.1002/jee.20349. 25

Elvezio, C., Sukan, M., Oda, O., Feiner, S., and Tversky, B. (2017). *ACM SIGGRAPH 2017 VR Village*, 1–2. DOI: 10.1145/3089269.3089281. 62

Fahmi, F., Tanjung, K., Nainggolan, F., Siregar, B., Mubarakah, N., and Sarlis, M. (2020). Comparison study of user experience between virtual reality controllers, leap motion controllers, and senso glove for anatomy learning systems in a virtual reality environment, *IOP Conference Series Materials Science and Engineering,* 851:012024. 50

Feldman, D. P. (2019). *Chaos and Dynamic Systems.* Princeton University Press. 13

Ferster, B. (2013). *Interactive Visualization: Insight through Inquiry.* MIT Press.

Freeman, S., Eddy, S, L., McDonough, M., Smith, M. K., Okoroafor, N., Jordt, H., and Wenderoth, M. P. (2014). Active learning increases student performance in science, engineering, and mathematics. *Proceedings of the National Academy of Sciences of the United States of America.* 111(23), 8410–8415. DOI: 10.1073/pnas.1319030111. 29

Feil, A. and Mestre, J. P. (2010). Change blindness as a means of studying expertise in physics. *Journal of the Learning Sciences,* 19(4), 480-505. DOI: 10.1080/10508406.2010.505139. 35

Goodnough, K. (2007). Enhancing pedagogical content knowledge through self-study: an exploration of problem-based learning. *Teaching in Higher Education,* 11(3). DOI: 10.1080/13562510600680715. 29

Gupta, A., Hammer, D., and Redish, E. F. (2010). The case for dynamic models of learners' ontologies in physics. *Journal of the Learning Sciences,* 19(3), 285–321. DOI: 10.1080/10508406.2010.491751. 34, 71

Haghanikar, M. M. (2012). Exploring students' patterns of reasoning. Thesis. Kansas State University. Manhattan, KS. 30, 32, 33

Haghanikar., M. M. (2019). *Cyberlearning and Augmented Reality in STEM Education.* Media Conference (GEM). 46

Haghanikar, M. M. (2020). Designing virtual tools to teach science post-pandemic. *Nature Volve,* 6.

He, Z., Wu, L., and Li, X. (2018). When art meets tech: The role of augmented reality in enhancing museum experiences and purchase intentions. *Tourism Management,* 68, 127-139. DOI: 10.1016/j.tourman.2018.03.003. 55

Hestenes, D. (1987). Toward a modeling theory of physics instruction. *American Journal Physics,* 55(5), 440–454. DOI: 10.1119/1.15129. 29

Hake, R. R. (1992). Socratic pedagogy in the introductory physics lab. *Physics Teacher,* 30, 546–552. DOI: 10.1119/1.2343637.

Harada, A. and Takashima, Y. (2012). *Supramolecular Chemistry: Cyclodextrin-Based Supramolecular Polymers* (A. Harada, Ed.). DOI: 10.1002/9783527639786.

Hegarty, M. and Sims, V. K. (1994). Individual differences in mental animation during mechanical reasoning. *Memory and Cognition,* 22(4), 411–430. DOI: 10.3758/BF03200867. 26

Hegarty, M. and Waller, D. A. (2005). Individual differences in spatial abilities. In P. Shah (Ed.) and A. Miyake, The Cambridge Handbook of Visuospatial Thinking (p. 121–169). Cambridge University Press. DOI: 10.1017/CBO9780511610448.005. 24

Hegarty, M., Montello, D. R., Richardson, A. E., Ishikawa, T., and Lovelace, K. (2006). Spatial abilities at different scales: Individual differences in aptitude-test performance and spatial-layout learning. *Intelligence*, 34(2) 151–176. DOI: 10.1016/j.intell.2005.09.005. 24

Hegarty, M. (2014). Spatial thinking in undergraduate science education. *Spatial Cognition and Computation*, 14(2). DOI: 10.1080/13875868.2014.889696. 23, 24, 25, 26, 34

Hegarty, M., Stieff, M., and Dixon, B. (2015). Reasoning with diagrams: Towards a broad ontology of spatial thinking strategies. In *Space in mind: Concepts for Spatial Learning and Education* (pp. 75–98). Cambridge, MA: MIT Press. 25

Hernández, N. E., Hansen, W. A., Zhu, D., Shea, M. E., Khalid, M., Manichev, V., Putnins, M., Chen, M., Dodge, A. G., Yang, L., Murrero-Berrios, I., Banal, M., Rechani, P., Gustafsson, T., Feldman, L. C., Lee, S-H., Wackett, L. P., Dai, W., and Khare, S. D. (2019). Stimulus-responsive self-assembly of protein-based fractals by computational design. *Nature Chemistry*, 11, 605–614. DOI: 10.1038/s41557-019-0277-y. 16

Herrera, F., Bailenson, J., Weisz E., Ogle, E., and Zaki, J. (2018). Building long-term empathy: A large- scale comparison of traditional and virtual reality perspective-taking. *PLoS ONE*, 13(10). DOI: 10.1371/journal.pone.0204494.

Hilgenfeldt, S. and Schulz, H.-C. (1994). Experimente zum Chaos. *Specktrum der Wissenschaft* 72(1). 10

Hou, L. and Wang, X. (2013). A study on the benefits of augmented reality in retaining working memory in assembly tasks. A focus on differences in gender. In: *Automation in Construction* 32, S. 38–45. DOI: 10.1016/j.autcon.2012.12.007. 65

Janelle, D. G., Hegarty, M., and Newcombe, N. S. (2014). Spatial thinking across the college curriculum: A report on a specialist meeting. *Spatial Cognition and Computation*, 14(2), 124–141. DOI: 10.1080/13875868.2014.888558. 25

Johnstone, A. H. (1991). Why is science difficult to learn? Things are seldom what they seem. *Journal of Computer Assisted Learning*, 7(2), 75–83. DOI: 10.1111/j.1365-2729.1991.tb00230.x. 36

Karplus, R. (1974). The learning cycle. In *The SCIS Teachers Handbook*. Berkeley, CA: Regents of the University of California. 44

Kenkel, N. C. and Walker, D. J. (1993). Fractals and ecology. *Abstrata Botanica*, 17(1–2), 53–70.

Kenkel, N. C. and Walker, D. J. (1996). Fractals in the Biological Sciences. *Coenoses* 11, no. 2 : 77–100. Accessed May 20, 2021, http://www.jstor.org/stable/43461170. 17

Kirschner, P. A., Sweller, J., and Clark, R. E. (2006). Why minimal guidance during instruction does not work: An analysis of the failure of constructivist, discovery, problem-based, experiential, and inquiry-based teaching, *Educational Psychologist*, 41:2, 75-86, DOI: 10.1207/s15326985ep4102_1. 70

Kozhevnikov, M., Motes, M. A., and Hegarty, M. (2007). Spatial visualization in physics problem solving. *Cognitive Science*, 31, 549–579. DOI: 10.1080/15326900701399897. 25

Kress, G. (2009). *Multimodality: A Social Semiotic Approach to Contemporary Communication* (1st ed.). Routledge. DOI: 10.4324/9780203970034. 73

Lawson, A. E. and Thompson, L. D. (1988). Formal reasoning ability and misconceptions concerning genetics and natural selection. *Journal of Research in Science Teaching*, 25(9), 733–746. DOI: 10.1002/tea.3660250904. 31

Lawson, A. E., Alkhoury, S., Benford, R., Clark, B. R., and Falconer, K. A. (2000). What kinds of scientific concepts exist? Concept construction and intellectual development in college biology. *Journal of Research in Science Teaching*, 37(9), 996–1018. DOI: /10.1002/1098-2736(200011)37:9<996::AID-TEA8>3.0.CO;2-J. 30

Lee, S. A., Chung, A. M., Cira, N., and Riedel-Kruse, I. H. (2015). Tangible interactive microbiology for informal science education. *TEI '15: Proceedings of the Ninth International Conference on Tangible, Embedded, and Embodied Interaction*. DOI: 10.1145/2677199.2680561. 45

Lindgren, R., Tscholl, M., Wang, S., and Johnson, E. (2016). Enhancing learning and engagement through embodied interaction within a mixed reality simulation. Computers and Education, 95, 174-187. DOI: 10.1016/j.compedu.2016.01.001. 35

Manderlbrot, B. B. (1982). *The Fractal Geometry of Nature.* Time Books. 7

Margaryan, A., Bianco, M., and Littlejohn, A. (2015). Instructional quality of massive open online courses (MOOCs). *Computer and Education*, 80, 77–83. DOI: 10.1016/j.compedu.2014.08.005. 39

Mason, B. (2012). Collaborating to improve science teaching and learning through the ComPADRE digital library I. *Journal Acoustical Society of America* 132(1921). DOI: 10.1121/1.4755051. xxii

McBride, D. L., Zollman, D., and Rebello, N.S. (2010). Method for analyzing students' utilization of prior physics learning in new contexts. *Physical Review, Special Topics: Physics Education Research*, 6(6). 30

McDermott, L. C. and Shaffer, P. S. (1992). Research as a guide for curriculum development: An example from introductory electricity. Part I: Investigation of student understanding, *American Journal of Physics*, 60 , 994–1003. DOI: 10.1119/1.17003. 36, 57

McDermott, L. (1996). *Physics By Inquiry*. John Wiley and Sons.

McGoodwin, M. (2000). Julia jewels: An exploration of Julia sets. Retrieved April 5, 2021 from http://www.mcgoodwin.net/julia/juliajewels.html. 7

McGrath, D., Savage, C., Williamson, M., Wegener, M., and McIntyre, T. (2008). Teaching special relativity using virtual reality. In A. Hugman and K. Placing (Eds.) *Symposium Proceedings: Visualization and Concept Development.*

Mermin, N. D. (2005). It's about time: Understanding Einstein's relativity. Princeton University Press. *JSTOR*, www.jstor.org/stable/j.ctt7t136. Accessed 20 May 2021. 62

Moore, K., Giannini, J., and Losert, W. (2014). Toward better physics labs for future biologists. *American Journal of Physics*, 82(387). DOI: 10.1119/1.4870388. 45

Moreau, D. (2012). The role of motor processes in three-dimensional mental rotation: Shaping cognitive processing via sensorimotor experience, *Learning and Individual Differences*, Volume 22(3) 354–359. DOI: 10.1016/j.lindif.2012.02.003. 55

Morphew, J. ,Mestre, J., Ross, B., and Strand, N. (2015) Do experts and novices direct attention differently in examining physics diagrams? A study of change detection using the flicker technique, *Physical Review Physics Education Research*, 11, 020104. DOI: 10.1103/PhysRevSTPER.11.020104. 35

Muikku, J. and Kalli, S. (2017). The IMD Project. VR/AR Market Report. http://www.digitalmedia.fi/wp-content/uploads/2018/02/DMF_VR_report_edit_180124.pdf. 47

Nae, H.-J. (2017). An interdisciplinary design education framework. *Design Journal*, 20(1) S835–S847. DOI: 10.1080/14606925.2017.1353030. 1

Nature Communications. (2019). *Supramolecular Chemistry*. https://www.nature.com/collections/wypqwypccc.

Nieswandt, M. and Bellomo, K. (2009). Written extended-response questions as classroom assessment tools for meaningful understanding of evolutionary theory. *Journal Research in Science Teaching*, 46(3), 333–356. DOI: 10.1002/tea.20271. 31, 32

NSTA. https://www.nsta.org/science-standards.

Ong, S. (2017). *Beginning Windows Mixed Reality Programming*. Apress. 25, 51

Organtini, G. (2018). Arduino as a tool for physics experiments. *Journal of Physics: Conference Series*, DOI: 10.1088/1742-6596/1076/1/012026. 46

Orion, N., Ben-Chaim, D., and Kali, Y. (1997). Relationship between earth-science education and spatial visualization. *Journal of Geoscience Education*, 45(2), 129-132. DOI: 10.5408/1089-9995-45.2.129. 25

Palmer, T. (2008). Edward Norton Lorenz. *Physics Today*, 61(9). DOI: 10.1063/1.2982132. 16, 17

Pell, M. (2017). *Envisioning Holograms, Design Breakthrough Experiences for Mixed Reality*. Apress. DOI: 10.1007/978-1-4842-2749-7. 55

Piaget, J. and Inhelder, B. (1969). *The Psychology of the Child*. Basic Books.

Redish, E. F. (1994). The implications of cognitive studies for teaching physics, *American Journal of Physics*. v62(6), p796–803. 35

Redish, J. (2014). NEXUS/Physics: An interdisciplinary repurposing of physics for biologists. *American Journal of Physics*, 82(368). DOI: 10.1119/1.4870386. 45

Reid, N. and Yang, M-J. (2002). The solving of problems in chemistry: The more open-ended problems. *Research in Science and Technological Education*, 20(1), 83-98. https://www.learntechlib.org/p/165517/. 36

Reuell, P. (2012). 3-D image shows how DNA packs itself into a fractal globule. *Harvard Gazzette*. https://scitechdaily.com/3-d-image-shows-how-dna-packs-itself-into-a-fractal-globule/. 16

Russ, R. S., Scherr, R. E., Hammer, D., and Mikeska, J. (2008). Recognizing mechanistic reasoning in student scientific inquiry: A framework for discourse analysis developed from philosophy of science. *Science Education*, 92(3), 499-525. DOI: 10.1002/sce.20264. 71

Sanger, M. J. and Greenbowe, T. J. (2000). Addressing student misconceptions concerning electron flow in aqueous solutions with instruction including computer animations and conceptual change strategies, *International Journal of Science Education*, 22:5, 521-537, DOI: 10.1080/095006900289769. 42

Scheri, III, S. R. (2005). *The Casino's Most Valuable Chip: How Technology Transformed the Gaming Industry*. Institute for the History of Technology.

Scherr, R., Shaffer, P. S., and Vokos, S. (2001). Student understanding of time in special relativity: Simultaneity and reference frames. *American Journal of Physics*, 69(S24). DOI: 10.1119/1.1371254. 42

Schmidt, H. W. (2012). *Supramolecular Polymer Chemistry*. A. Harada (Ed.). John Wiley and Sons, Inc. DOI: 10.1002/anie.201206234. 17

Schaad, L., Hlushchuk, R., Barré, S., Gianni-Barrera, R., Haberthur, D., Banfi, A., and Djonov, V. (2017). Correlative imaging of the murine hind limb vasculature and muscle tissue by microct and light microscopy. *Scientific Reports*, 7, 41842. DOI: 10.1038/srep41842. 73

Sedikidies, C. and Skoronski, J. J. (1991). The law of cognitive structure activation. *Psychology Inquiry*, 2, 169-184. DOI: 10.1207/s15327965pli0202_18. 3

Segré, D., Ben-Eli, D., Deamer, D. W. and Lancet, D. (2001). The lipid world. *Origins Life and Evolution of the Biosphere*, 31, 119-145. DOI: 10.1023/A:1006746807104. 17

Seo, S.-W., Kwon, S., Hassan, W., Talhan, A., and Jeon, S. (2019). Interactive virtual-reality fire extinguisher with haptic feedback. *VRST '19: 25th ACM Symposium on Virtual Reality Software and Technology*. DOI: 10.1145/3359996.3364725. 50

Sethian, J. A. and Strain, J. (1992). Crystal growth and dendric solidifciation. *Journal Computational Physics*, 98(2), 231-253. DOI: 10.1016/0021-9991(92)90140-T. 14

Shen, Y., R. and Yang, G.-Z. (2016). Theory of self-phase modulation and spectral broadening. In *The Supercontinuum Laser Source*, Chapter 1. Provided by the SAO/NASA Astrophysics Data System. DOI: 10.1007/978-1-4939-3326-6_1. 72

Slotta, J. D. and Chi, M. T. H. (2006). Helping students understand challenging topics in science through ontology training. *Cognition and Instruction*, 24, 261–289. DOI: 10.1207/s1532690xci2402_3. 34

Sokoloff, D. R. and Thornton, R. K. (1997). Using interactive lecture demonstrations to create an active learning environment, *Physics Teacher*, 35 340. DOI: 10.1119/1.2344715. 68

Soler J. L., Contero, M., and Alcañiz, M. (2017). VR serious game design based on embodied cognition theory. *Conference: Joint International Conference on Serious Games*, 12–21. DOI: 10.1007/978-3-319-70111-0_2. 35

Symons, J. and Boschetti, F. (2013). How computational models predict the behavior of complex systems. *Foundations Science*, 18, 809–821. DOI: 10.1007/s10699-012-9307-6. 5

Tolentino, N. A. and Roleda, L. S. (2019). Gamified physics instruction in a reformatory classroom context. *IC4E '19: Proceedings of the 10th International Conference on E-Education, E-Business, E-Management and E-Learning*, 135–140. DOI: 10.1145/3306500.3306527. 42

Thornton, R. K. (1987). Tools for scientific thinking—microcomputer-based laboratories for teaching physics. *Physics Education*, 22, 230–238. DOI: 10.1088/0031-9120/22/4/005. 40

Vygotsky, L. S. (1978). *Mind in Society: The Development of Higher Psychological Processes*. Cole, M., John-Steiner V., Scribner S., and Souberman E. (Eds.), Harvard University Press.

Wai, J., Lubinski, D., and Benbow, C. P. (2009). Spatial ability for STEM domains: Aligning over 50 years of cumulative psychological knowledge solidifies its importance. *Journal of Educational Psychology*, 101(4), 817–835. DOI: 10.1037/a0016127. 24

Wegener, M. J., McIntyre, T. J., McGrath, D., and Williamson, M. (2012). Developing a virtual physics world. *Australasian Journal of Educational Technology*, 28(3), 504–521. DOI: 10.14742/ajet.847. 43, 62

Wess Institute (2016). *Self-Assembly: Nature's Design Principle.* https://wyss.harvard.edu/news/self-assembly-natures-design-principle/.

Wittman, M. C. (2002). The object coordination class applied to wave pulses: Analyzing student reasoning in wave physics. *International Journal of Science Education*, 24(1). DOI: 10.1080/09500690110066944. 34, 71

Wolfram, S. (2020). *A Project to Find the Fundamental Theory of Physics.* Wolfram Media, Inc. 2

Wu, H.-K. and Shah, P. (2004), Exploring visuospatial thinking in chemistry learning. Sci. Ed., 88: 465-492. DOI: 10.1002/sce.10126. 36

Zhu, G. and Singh, C. (2009). Students' understanding of stern gerlach experiment. *AIP Conference.* DOI: 10.1063/1.3266744. 58

Zhu, G. and Singh, C. (2012). Surveying students' understanding of quantum mechanics in one spatial dimension. *American Journal of Physics*, 80(252). DOI: 10.1119/1.3677653. 58

Zollman, D., Rebello, S., and Hogg, K. (2003). Quantum mechanics for everyone: Hands-on activities integrated with technology. *American Journal of Physics*, 70(252). DOI: 10.1119/1.1435347. 44

Author Biography

Mojgan Haghanikar, Ph.D., Physics and Astronomy Education, is a physics educator and Ed-Tech influencer for science and engineering education. Holding a degree in solid-state physics, she started her early career in a laser lab, creating traditional holograms for industrial applications, directing outreach programs, and developing science museums. In 2002, she received a CHEVE-NING scholarship from the British Foreign and Commonwealth office that allowed her to complete a Master of Science in astronomy education at the University of Glasgow. Soon after, she was admitted to the Kansas State University Department of Physics, where she concluded her academic ventures with a Doctor of Philosophy in physics education.

Haghanikar's doctoral dissertation involved data analytics that modeled undergraduate students' scientific reasoning patterns through cognitive lenses. The methodology of her research was qualitative and quantitative multi-stages cluster analysis. The outcome was statistical logistic regression models that described students' higher levels of thinking processes regarding the various modes of instruction. The cluster analysis modeling revealed that the lack of visibility of dynamic systems' interrelationships in various scales and time spans strongly affected students' knowledge structure quality. Therefore, she was motivated to design various demonstrations and lab activities to enhance dynamic systems' visibility. She started with inventing experimental projects and demonstrations that helped students' visualization. Although the demonstrations were practical, the resources were limited, and it was not possible to replicate designs in various locations. With the advent of emerging technologies, Haghanikar replicated the designs using mixed reality platforms to enhance scientific demos and labs' availability and scalability.

Since 2011, Haghanikar has served in several roles, such as visiting assistant professor and lecturer at California Polytechnic State University and Towson University, to name a few. She was an educational designer and research scientist for nanotech engineering and industrial education at SUNY Polytechnic. She also served as an educational designer for physics and astronomy online courses and further created simulation design and game design for emerging technology platforms. Haghanikar mentored subject matter experts to incorporate appealing features to their courses such as simulations, self-diagnostic assessment with adaptive pace, animations, interactive games, and engaging tutorials.

As a science education research affiliate, she collaborates with SETI Institute serving as PI on several grants to use robotic telescopes and emerging technologies to promote cyberlearning in teaching physics, astronomy, and quantum computing.